Essays on Forensic Engineering

N. Krishnamurthy

Safety Consultant and Trainer
www.profkrishna.com
Singapore

ALSO BY THE SAME AUTHOR:
***INTRODUCTION TO ENTERPRISE
RISK MANAGEMENT***
ALSO THROUGH AMAZON
BROWSE WITH KEYWORD: 1539436284

ISBN: 197397004X
ISBN-13: 978-1973970040

DEDICATION

To the many professionals in USA and Singapore who involved me in their forensic efforts, and to the conference organizers in India who gave me a platform to share my experiences and opinions.

ACKNOWLEDGMENTS

Author expresses his deep gratitude to the following persons and their organizations:

Mr. R. Sundaram
President, Structural Engineers World Congress, India

- For SEWC inviting me to present a paper at the World Congress at Bangalore in 2007, and giving me permission to reprint it in this book.

Mr. B.S.C. Rao
President, Association of Consulting Civil Engineers (India), 2013
Chairman, Programme Committee, ACCE(I)

- For ACCE(I) inviting me to present papers at the First and Second International Conferences on Forensic Civil Engineering at Bangalore in 2013 and at Nagpur in 2016.

Mr. Ajit Sabnis
President, Association of Consulting Civil Engineers (India), 2016

- For ACCE(I) giving me permission to reprint my three papers from the (First) Conference on Forensic Civil Engineering at Bangalore in 2013, and my two papers from the Second Conference on FCE at Nagpur in 2016 in this book.

———

CONTENTS

PREFACE

Merriam-Webster Dictionary defines 'Essay' as follows:
Noun: A short piece of writing that tells a person's thoughts
or opinions about a subject
Verb: To try to do, perform, or deal with (something)

It is in both these senses that I offer this collection of six published papers of mine on accident investigation and prevention in engineering-related situations, which broadly fall under the term 'forensic engineering'. I have taken the liberty of including cases which were not mishaps but were intended to prevent mishaps before they became court cases as 'pro-active' forensic engineering.

My forays into this exotic area have been few and far between, but (to me) quite interesting and challenging, to the extent that even though my 'achievements' are miniscule when compared to even the average professional forensic engineer, I felt like sharing my experiences with a wider audience. My efforts are strictly amateurish and my impact on the profession in this regard quite regional and limited. But hey, I learnt a lot and I hope my readers may also get a kick out of it.

I feel lucky I got started on digital computing as early as 1959 in the USA and have kept up with the rapidly expanding field – particularly in computer graphics – as best as I could, because all of these came in quite handy in my accident investigations as skills complementing my structural engineering specialty.

I did not plan to be a forensic engineer or expert witness. They came by me – or I by them – by happenstance. Some senior official would want me to investigate an accident involving complex analysis; some lawyer would want me to see if I can get his worker client a better settlement; sometimes, that I was a 'Lone Ranger' with no obligation to toe any party line or company policy would drive someone to seek my help in a messy PR situation. I would take them, only refusing when something went against my grain.

Many cases I accepted were settled out of court. In those that were litigated, my clients did not always win. But then, forensic engineers and expert witnesses do not win or lose; they are just expected to argue for the truth as they see it based on their knowledge and experience.

Since I have personally handled only a limited number of cases, when time came for me teach engineering ethics in two universities, I got into studying technological failures in depth and accumulated a lot of borrowed experience, mostly available from USA, where I had studied and worked while many of the cases discussed actually happened – so that gave me a third dimension to the entire pursuit.

I offered a few short courses in forensic engineering to local engineers; I am currently teaching a course for a local university in which this subject is a module. These, together with some very unusual cases I investigated, led to the papers in this book.

For legal reasons, forensic engineer cannot discuss their cases freely or in detail. I have had to control my impulse to be frank and transparent in my talks and writings, to the extent that I had to even substitute an equivalent problem to the actual one for fear that the actual situation would trigger identification.

In my papers, I have freely borrowed from the vast forensic engineering literature available in libraries and websites. As almost all my extraneous examples are in the public domain, I have not felt the need to acquire permission to present the facts or pictures.

I have scrupulously cited the references from which I gathered specific material for every paper. If I have omitted any author or copyright owner, it would be by oversight or ignorance and not out of negligence. If the person or organization points out where my omission is and how I should cite the source, I will be happy to rectify the omission.

Some of the internet links listed in the References may be broken or missing. That too is in the nature of things. But readers can always find that or similar material from the web by inserting some keywords, or even the obsolete URL citation into the search box of the browser, and more often than not, find what they need.

Copyright laws are funny: I casually wrote to a local newspaper asking how I should cite a quote and show a picture of their coverage of my testimony in a case, and back came the reply that I could do so after paying a three-figure fee. But they had not asked my permission or paid a fee when they printed my words and picture! I simply cut their material out from my papers.

Some 'classical' cases such as the Regency Hyatt Walkway collapse and NASA shuttle disasters have been mentioned in more than one of my papers. Although it was necessary for me to briefly describe the mishap each time, I have treated each discussion differently from the others to focus on the theme of the paper.

I have made make certain non-substantive changes from the original versions. First, as I made the paper number as the chapter number, this number was added as prefix to the Section and Sub-section numbers, figure and table numbers of the individual essays. Second, all references were modified to approximately the same format. [In the text, references will be uniformly listed within square brackets.] Finally, I deleted a few names and pictures to further reduce chances of identifying the cases which were not yet in public domain. I added a figure in place of two I removed. I cut details of one case. I hope readers will accept the need for these.

Because of the nature of the beast, there is a lot of subjective — even biased — opinions expressed in these pages. I hope readers accept them in the spirit in which I offer them, namely, purely my desire and enthusiasm to share what I have found and learnt on a very difficult topic, subject to revision and correction as and when someone points out the error of my ways. My pleasure lies in that I stand at least in the same line of service to the profession as stalwarts in the field do.

My goal in publishing these papers will be fulfilled if the reader is informed and guided, or at least entertained by my effort!

Happy reading!

N. Krishnamurthy   *July 2017*

THE COVER

The author's design of the cover highlights his symbolic concept of the arduous journey from a mishap through site investigation, micro and macro testing and complex mathematical and theoretical analysis to final professional and legal resolution ultimately to benefit society worldwide.

1

FORENSIC ENGINEERING IN STRUCTURAL DESIGN AND CONSTRUCTION†

ABSTRACT

When a structure fails, there is invariably an investigation to find out why it failed. Apart from the legal and professional necessity to determine the cause of failure, there is also the need to learn from it lessons that would enable subsequent designers and builders or fabricators to avoid the pitfalls of the failed structure and develop safer alternatives.

Technological developments in recent decades have introduced new configurations, materials and methods of design and construction that raise new and complex problems. Failures are caused by many unprecedented causes singly or in combination. Paradoxically, in this pursuit of novelty and innovation, even basic principles of sound structural design and good construction practice are often violated, leading to failures.

Author has more than five decades of teaching, research, and consulting experience in structural engineering, computer applications, and recently, in construction safety, failure investigation, and risk management, in USA, Singapore, and India. During his professional service, he has had many opportunities to study structural and construction failures, discuss with other professionals, collaborate with forensic experts, and also report and testify on failure investigations.

This paper will review the scope of forensic engineering in the light of the increased need for increased attention to and proper documentation of the design/construction process; discuss the preparation required for failure investigation; point out critical procedures and protocols in the actual investigation; and illustrate its application by a few classical case studies from around the world, and a few from his own experience.

1.1 FORENSIC ENGINEERING BASICS

1.1.1 What is forensic engineering?

"Forensis" means 'Public' in Latin; 'forensic' has come to refer to legally sustainable documentation, usually applied to accidents, crimes, etc.

In particular, forensic engineering is the application of the art and science of engineering in the jurisprudence (legal) system, requiring the services of qualified experts. [1.1.]

[All reference citations in this book will be numbers in square brackets.]

Forensic engineering may include the investigation of the physical causes of accidents and other sources of claims and litigation, preparation of engineering reports, testimony at hearings and trials in administrative or judicial proceedings, and rendition of advisory opinions to assist the resolution of disputes affecting life or property.

Generally the purpose of a forensic engineering investigation is to determine cause or causes of failure with a view to improve performance or life of a component, or to assist a court in determining the facts of an accident.

1.1.2 Who can be a forensic engineer?

Anyone who is an expert in the subject under investigation

- Has the necessary formal education
- Has the necessary experience
- Is licensed, or is otherwise recognised as an expert
- Is active in technical societies

And who is fair, impartial, and ethical

- Is truthful
- Is objective
- Avoids conflict of interest
- Focuses on analysis, design, and technology rather than on fixing blame on persons responsible.

The author believes that a forensic engineer needs further a special mindset to be able to carry out such an assignment. [1.2.] To track beyond the obvious, and find the root cause behind the immediate cause of failure, he (– the male pronoun will cover the female equivalent also, unless gender specific, hereinafter) must:

- "Open his third eye" and not only see more, but also hear more, smell more, taste more, and feel more than the average person.
- Understand more than others about what is going on around him, look under the carpet and behind the screen, and read between the lines.

1.2 TYPES OF FAILURE

Failure need not always mean that a structure collapses. It can make a structure deficient or dysfunctional in usage. It may even cause secondary adverse effects.

(a) Safety failure – Injury, death, or even risk to people:
- Collapse of formwork during concrete placement
- Punching shear failure in flat slab concrete floor
- Trench collapse
- Slip and fall on wet floor

(b) Functional failure – Compromise of intended usage:

- Excessive vibration of floor
- Roof leaks
- Inadequate air conditioning
- Poor acoustics

(c) Ancillary failure – Adverse effect on schedules, cost, or use:
- Delayed construction
- Unexpected foundation problems
- Unavailability of materials
- Strikes, natural disasters, etc.

Matousek and Schneider [1.3] cite the figures in Table 1.1, for sudden failures (a) and unacceptable conditions (b and c above).

Table 1.1 Sudden Failures and Unacceptable Conditions

Category	Type of failure/damage	%
Sudden failures **Subtotal = 63%**	Loss of equilibrium	13
	Failure with collapse	29
	Failure without collapse	11
	Other types of failures	10
Unacceptable conditions **Subtotal = 37%**	Excessive cracks	16
	Deflections and change of shape	7
	Errors in dimensions and support conditions	8
	Other unacceptable conditions	6
	Total	**100**

1.3 SOURCES AND CAUSES OF FAILURE

1.3.1 Errors and stages leading to failures

Matousek and Schneider [1.3] also cite reasons for failure in percentages, as follows:

- Ignorance, carelessness, negligence   35%
- Forgetfulness, errors, mistakes   9%
- Reliance upon others without sufficient control   6%
- Underestimation of influences   13%
- Insufficient knowledge   25%

- Objectively unknown situations (unimagined?)   4%
- Others   8%

Table 1.2 lists distribution of failure by stages of its occurrence, according to various authors.

Table 1.2 Distribution of Failures over Various Stages

Reference	P&D	C	U&M	OMF	T
Matousek	37	35	5	25	100
Brand & Glatz	40	40	-	20	100
Yamamoto & Ang	36	43	21	-	100
Grunau	40	29	*31	-	100
Reygaertz	49	22	*29	-	100
Melchers, et al	55	24	21	-	100
Fraczek	55	53	-	-	^108
Allen	55	49	-	-	^104
Hadipriono	19	27	33	20	99
Average	*43*	*36*	*16*	*7*	*^102*

P&D = Planning and Design, C = Construction, U&M = Use and Maintenance, OMF=Others and Multiple Factors, T = Total. All numbers percentages. * = Includes materials, environment, and service conditions. ^ = Multiple effects for single cause, >100

1.3.2   Type of errors in design/planning

In 295 cases of damaged structures, the types of errors in design and planning were as follows:

- Conceptual errors   34%
- Structural analysis   34%
- Drawings and specifications   19%
- Work planning and preparation   9%
- Combinations   4%

1.3.3   Primary causes of failure and how handled

Walker (1981) has summarised findings by numerous authors as shown in Table 1.3. In 723 cases of damaged structures, the damage-initiating influence was taken into account in the building process as follows:

- No consideration   26%
- Incorrect consideration   26%
- Insufficient consideration   16%
- Considered but risk accepted   22%
- Consideration unknown   10%

Table 1.3 Primary Causes of Failure

(a) Can be remedied by increased safety factors in structural design; unfavourable random effects lead to failure. %		
Inadequate appreciation of loading conditions or real behaviour of structure	36	
Inadequate appreciation of loadings or real behaviour of connections	7	
Excessive reliance on construction accuracy	2	
Serious mistakes in calculations and drawings	7	
Inadequate information in contract documents and instruction	4	
Contravention of requirements in contract documents and instruction	9	
Inadequate execution of erection procedure	13	
Unforeseen misuse, abuse and/or sabotage, natural catastrophe, deterioration	7	
Others	5	
Sub-total for (a)		**90**
(b) Cannot be counteracted or avoided by increased safety factors in structural design; gross human errors reducible by checking and supervision. %		
Unfavourable load variation or combination (related to partial factors for loads), present but small	0	
Inaccuracies in design assumptions of support conditions, hinges etc, (related to model uncertainties)	3	
Material and workmanship deficiencies (related to partial resistance factors)	4	
Foreseeable deterioration	3	
Sub-total for (b)		**10**
Total		**100**

1.3.4   Risk probabilities

Table 1.4 lists the risk probabilities for various common activities, based on American data. The death rate in number of deaths per hour times the exposure of the person performing the activity in number of hours per year, gives the death risk in number of deaths per year.

Table 1.4 Risk Probabilities for Various Activities

Activity [* Estimated average per person]	Death rate, No. per hr. of exposure (10^{-9})	Exposure, No. of hrs. per year	Death risk, No. per year (10^{-6})
Alpine climbing	30,000 to 40,000	50	1500 to 2000
Cigarette smoking	2500	400	1000
Coal mining (UK)	210	1500	300
Construction work	70 to 200	2200	150 to 440
Car travel	700	300	200
Swimming	3500	50	170
Boating	1500	80	120
Manufacturing*	20	2000	40
Air travel	1200	20	24
Building fires*	1 to 3	8000	8 to 24
Train travel	80	200	15
Structural failures	0.02	6000	0.1

The table clearly shows that (permanent) structural failures are certainly not the disastrous killer that many may consider them to be. On the other hand, construction work involving temporary structures is 1500 to 4400 times as risky, and demands considerable attention from forensic engineers.

1.4 PRACTICE OF FORENSIC ENGINEERING

1.4.1 Forensic engineer's kit

The kit should contain some or all of the following items, depending on the size and complexity of the accident:

- The LENS, of course
- Graph & Lined Paper
- Pencils/Pens
- Flashlight w/batteries
- Small Blackboard w/chalk *
- Ruler and Tape Measure

- Warning Signs
- Camera and Film **
- Sample Containers
- Safety helmet, gloves, shoes
- Cassette Recorder w/tape
- Data Collection Forms
- High Visibility Marking Tape
- Calculator

[* – Blackboard and chalk today replaced by digital note-taking! – Author, June 2016]

** – Many courts do not admit digital photos as evidence, but today's forensic engineer must carry a digital camera as well as a video camera to record for later analysis. His skill set should include familiarity with computers, spreadsheets, and computer graphics.

1.4.2   When to start

The investigation should begin as soon as possible after the accident happens, as otherwise:

- **Operations may be disrupted.**
  The more serious an accident is, the more time and effort it takes to bring work back to normal. The sooner an investigation begins, the sooner normal operations can resume.
- **Memories may fade.**
  As time passes, what a person remembers can change. Interviewing witnesses as soon as possible after the accident helps assure a more accurate account of what happened.
- **Employees may be put to risk.**
  There is a good chance a similar accident can happen again unless the causes are identified and corrected. The earlier an investigator can determine the causes, the earlier corrective action can be taken to prevent a recurrence.

1.4.3   Sources of information

Gathering information from variety of sources helps to avoid overlooking possibly important information. To be thorough, an accident analysis should include information from all available and existing records including:

- Documents
- Newspaper reports
- Police reports and records
- Medical reports and records
- Testimony of officials and affected worker(s)
- Photographs, videos
- Physical evidence at the scene
- Statements of witnesses*

    * – Note that statement of witnesses must be carefully evaluated for consistency, credibility, bias, etc. It may also change at any stage of the proceedings!

1.4.4 Physical evidence

Physical evidence includes, but is not limited to, the following:

- Condition of site
- Condition of work environment
- Condition of machinery and equipment
- Condition of materials
- Permits to Work, Safe Work Procedures
- Supervisors' and signalmen's forms

    Permission in writing (or tape) must be obtained to:

- Enter site,
- Take photographs,
- Talk to personnel,
- View records at site,
- Conduct tests at site, and
- Take away samples from site

1.4.5 Chain of custody of evidence

To stand up in court, every bit of evidence collected from the site or other sources must be tracked and documented in detail from beginning to end. Hence, attention must be paid to:

- Location, witnesses present, weather conditions if relevant, number, dimensions, etc.;
- Each sheet of paper and each item of physical artefact to be uniquely identified, tagged, and listed;
- Written (signed) and/or photographic documentation of giver and receiver of evidence, with third party witness if possible;
- Facts as distinct from opinions and conjectures.

1.4.6 Boundaries and constraints

The investigator must:

- Limit himself to the field of his expertise;
- Support his conclusions with computations or other established precedents;
- Support his opinions with citations from published literature, or from universally accepted practice;
- Use only facts or corroborated opinions in conducting analyses and drawing conclusions;
- Develop all possible and credible failure scenarios;
- Analyse outcomes under various plausible situations, e.g. Different support or loading conditions for structure;
- Strictly avoid pre-conceived notions and prejudices, even when working for a private paying client;
- Document his credentials and references for all to see;
- Ensure that his theoretical and experimental analyses follow standard procedures and are reproducible;
- Think "out of the box", think laterally, remembering that most accidents are unique even though they may appear to fit a pattern;
- Accept that he too can be wrong in some part of his submissions;
- Be ready for twists and turns in the investigation as it progresses, as further expert evidence or new testimony is adduced; and,
- Be prepared to lose an argument (or the case itself!), and not

take it as a matter of prestige to win all the time.

1.4.7 Back analysis, Re-building, Simulation, and Re-enactment

Back analysis is a common technique of forensic investigation, involving the repetition of earlier computations, tests etc. with data extracted from later developments at site. This must not be used to justify an earlier conclusion, particularly if assumptions are involved; instead, the assumptions, and even the processes must be re-examined.

Re-building of a collapsed structure may be possible in certain situations, but it may be undertaken (even when economically and practically feasible) only if there are detailed original designs and plans from which it was built the first time, and not from memory.

Simulation is generally accomplished via the virtual world of the computer. In this, extreme care is necessary to reflect the actual conditions obtaining at the time and place of the failure.

Re-enactment of an accident is a very risky and high-profile activity, not to be undertaken lightly, and should only be used under the following conditions:

- When it can supply information that cannot be obtained in any other way.
- When it will aid in determining preventive action.
- When it is necessary to verify facts given by a witness or the injured employee.

If investigator decides to re-build or re-enact a process or scenario, he must make certain that the re-enactment does not result in a repetition of the injury or damage. Well-trained professionals (like certified stuntmen) may need to be employed with extra safeguards, to take the place of workers, injured or otherwise.

1.4.8 Report of investigation

The investigator's report normally includes the following items:

1. His qualifications/credentials as they pertain to the accident;

2. Terms of reference of assignment by the individual or organisation commissioning it;
3. Facts and brief review of the accident;
4. Summaries or reviews of existing analyses and reports by officials and witnesses;
5. Review and rebuttal of reports by opposing investigators and experts, etc. including points of concurrence;
6. Analyses of situations and events leading to the accident;
7. Possible credible scenarios for the accident;
8. Conclusions, prioritising the scenarios;
9. Recommendations for avoidance of similar accidents;
10. Statement of his contribution and disclaimers, if any.

1.4.9 Recommendations

Win or lose, and whether root causes for the accident can be found or not, the investigator should come up with improvements to the safety management system. Recommendations for deficiencies may be addressed under the following heads:

(a) Hazardous (unsafe) acts that occur because employees are:
- Unaware of the hazards they face and consequently do not know the special precautions that are necessary; or,
- Unable to perform in a safe and healthy manner because they are not properly trained, or they are not physically capable of performing the job; or because some aspect of the operation or work site prevents them from being able to work safely; or,
- Unmotivated to consider working safely as an important part of their job.

(b) Hazardous conditions that occur because they are either:
- Unnoticed: hazardous conditions that have not been identified during scheduled or informal inspections; or,
- Uncorrected: hazardous conditions that have been

identified, but have not been eliminated or controlled.

1.5 CASE STUDY 1 : DESIGN THEORY ERROR

1.5.1 Hyatt Regency Walkway Collapse

On July 17, 1981 in Kansas City, Missouri, USA, the walkway of Hyatt Regency Hotel opened the year before, on which hundreds of spectators were standing watching and enjoying the music and dance in the atrium area, collapsed, killing 114 people and injuring more than 200 others. [1.4, 1.5, 1.6.] At the time it was the deadliest structural collapse in U.S. history. (*See* Fig. 1.1 for before and after pictures of the collapse.)

The cause of the failure was traced to the failure of the flanges of two toe-to-toe welded channels at the suspension rod nut at the upper walkway, as shown by bold red arrows in the overall view and enlarged detail of Fig. 1.1. To simplify fabrication and erection, the contractor had replaced the original design of a single rod supporting both the suspended walkways, with two separate rods, one for each suspended walkway, the lower one being hung from the upper.

After all the exotic forensic theories and scenarios had been exhausted, it turned out that the disaster was caused by an overlooked fact of simple high school level statics: That the substitution of two rods in place of one in the particular fashion it was done doubled the load at the nut, as illustrated in Fig. 1.1.

While this turned out to be a one-line trivial solution by hand, a computer package would not have automatically caught it, unless the modelling had included the particular connection, modelled in considerable detail by finite elements – which is generally unduly expensive.

Apparently the original design was at least marginally safe. It was the fabrication change that halved the factor of safety. It appeared that the contractor had referred the change to the designer, but one of his assistants approved it after a cursory glance.

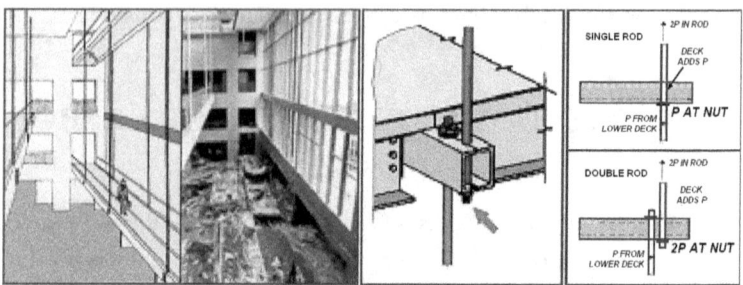

Fig. 1.1 (Left pair) Hyatt Regency Walkway, Before and After the Collapse; (Middle) Nut at Upper Walkway; (Right) Static Analysis of Single versus Double Hanger Rods

1.6 CASE STUDY 2 : PRE-FAB CONNECTION DESIGN FAILURE

1.6.1 Ronan Point Collapse, East London, UK, 16 May 1968

Ronan Point was a 23 storey apartment block in East London, assembled from pre-cast wall panels around floor slabs. Building started in 1966, and construction was completed on 11 March 1968. (Ref. 1.7, 1.8.) On 16 May 1968, a gas leak explosion on Floor 18 blew out kitchenette and living room walls. Same areas of higher Floors 19 to 22 collapsed, and the entire weight fell on the floor slab of Floor 18. Floors 17, 16, ... collapsed one by one, down up to the ground floor. 4 died and 17 were injured. (*See* Fig. 1.2.)

1.6.2 Findings

- Found no violation of any applicable building standard, nor defect in workmanship
- Building standards gave detailed requirements for design of elements, but little guidance on stability of entire structural system.

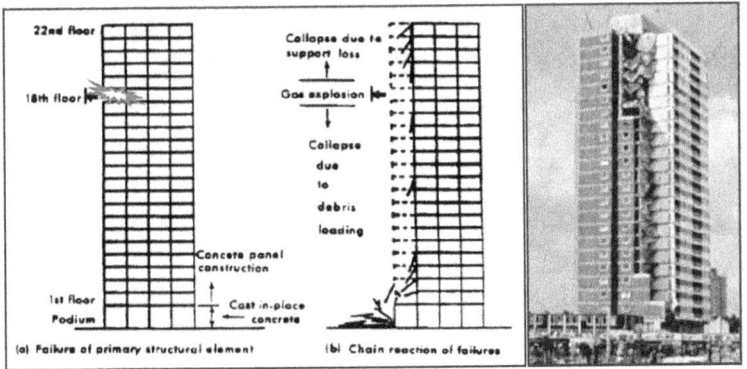

Fig. 1.2 Ronan Point: Schematic of Collapse, and Photograph

- Joint forces were resisted solely by bond, friction, and gravity.
- Upon removal of the walls, connections above could not redistribute loads since they were designed only for compression.
- Many connections were shoddily done.
- Existing building standards and codes of practice having general warnings and guidance on design of large panel structures to mitigate some of the problems were not consulted.

1.6.3   Root Causes and Response

Root causes were found to be:
- No structural redundancy
- Weak connection design
- Bad workmanship
The following actions were taken:
- All new buildings constructed after November 1968 and over 5 storeys were required to be able to resist an explosive force of 5 lbs per square inch.
- Existing buildings were allowed to resist an explosive force of 2.5 psi, provided that the gas supply was removed and flats were refitted for electric cooking and heating.
- In 1984, all nine blocks on the estate were demolished.
Most of the authoritative papers on progressive collapse were

published within a few years of the event. Interest in progressive collapse was immediately created in the United Kingdom and other nations, leading to changes to UK Building Regulations

- November 1968 – "Standards to Avoid Progressive Collapse – Large Panel Construction" – Had Alternate Load Path, Continuity, and Accidental Load
- April 1970 – Standards became mandatory
- 1974 – Provision of structural ties in British Standards

Ronan Point generated research and discussion in several countries. In the United States:

- 1972 – ANSI A58.1 addressed the issue
- 1976 – PCI included ties for pre-cast concrete walls
- Later events of 1970s influenced U.S. developments.

1.7 CASE STUDY 3 : DESIGN CHANGE FAILURE

1.7.1 Hartford Civic Center Arena Roof Collapse, 18 Jan. 1978

A 300' by 360' space frame for a stadium roof in Hartford, Connecticut, USA, completed on 16 Jan. 1973, collapsed on 18 Jan. 1978 under heavy water-ponding from a storm. [1.4, 1.5, 1.6.]

Luckily there were no spectators or other personnel in the stadium at the time. (*See* Fig. 1.3a for a sectional view, and Fig. 1.3d for a collapse scene.)

The space frame was an assembly of modular pyramidal pods, each 30' by 30', as shown in Fig. 1.3c.

1.7.2 Lapses in design and construction

Built in 1972-1973, the roof was an early example of a space truss, and of the use of computer-aided design. It was considered at the leading edge of design.

But certain code specifications were violated. Its use of the cruciform shape shown in Fig. 1.3b – a very inefficient form for compression loads – was itself a wrong start.

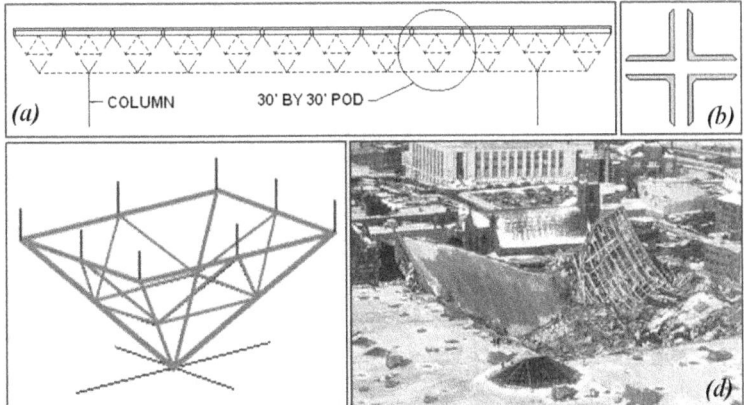

Fig. 1.3 Hartford Civic Center Arena: (a) Schematic Section; (b) Cruciform Section Used; (c) 30' by 30' pod; (d) Collapse Scene.

Warning signs were ignored during the erection process: The deflections at initiation of jacking up the roof were quite high, but overconfidence on the computer analysis made the designers ignore them, and instead urge the contractors to make ad-hoc arrangements to complete the erection without delay. Large deflections during normal use were also ignored.

1.7.3   Findings

Investigation showed that three design deficiencies responsible for collapse:
1. The top layer's exterior compression members on the east and the west faces were overloaded by 852%.
2. The top layer's exterior compression members on the north and the south faces were overloaded by 213%.
3. The top layer's interior compression members in the east-west direction were overloaded by 72%.

In addition to these design errors, there were bracing deficiencies as follows:
1. Midpoint braces for rods in the top layer were omitted.

2. The exterior rods were only braced every 30-feet, rather than the 15-feet intervals specified.
3. The interior rods were only partially and insufficiently braced at their midpoints.

1.7.4 Root Causes and Response

Major reasons for the reduction in design capacity were the changes in the connection configuration, as shown in Table 1.5.

Table 1.5 Comparison of (a) as designed, and (b) as-built connections

	Connection A	Connection B	Connection C	Connection D
(a) Original Design	Allowable force: 160,000-lb Allowable moment: 0	Allowable force: 185,000-lb	Allowable force: 625,000-lb	Allowable force: 565,000-lb
(b) As-built Design	Allowable force: 15,440-lb Allowable moment: 9,490 lb-ft	Allowable force: 59,000-lb	Allowable force: 363,000-lb	Allowable force: 565,000-lb

The key difference is that as-built diagonal members were attached some distance (only a few centimetres) below the horizontal members, thus unable to brace the horizontals against buckling.

The engineers for Hartford Arena depended on computer analysis to assess the safety of their design.

The roof design was extremely susceptible to buckling which mode of failure was not considered in that particular computer analysis and thus left undiscovered.

Incorporated into the computer model were some fundamental assumptions about end conditions on 30-ft. long members of the frame, grossly oversimplified. Connection details were difficult to incorporate in computer model. As a combination of these factors, and over-reliance on computer analysis with an imperfect model, the seriousness of the change in the connection was not revealed.

The collapse shook public confidence in space truss roofs. President Ford ordered water load testing for a similar roof over a museum in Michigan.

The failure tempered the tendency of engineers and architects to rely on computer models to cut down the structure to bare minimum, leaving no redundancy or margin for error.

1.7.5   Lessons learnt

1. Ensure that computer analysis/design covers ALL modes of failure. If it cannot, check out by conventional methods, those modes that cannot be (or would be too difficult to be) investigated by computers.
2. When any change is made to the design, reanalyse the as-proposed-to-be-built, with the eccentricities and changed support and joint conditions that actually would exist.
3. Heed warning signs during erection and construction. When deflections exceed computed values, stop the work, and evaluate the basic assumptions on which the computer analysis was performed. Do not start until the excessive deflection can be explained, and the problem corrected.

1.8 CASE STUDY 4 : BROAD REACH OF FORENSIC ENGINEERING

1.8.1   Northridge Earthquake - 1994

At 4:31 a.m. PST on Monday, Jan. 17, 1994, the ground shook for approximately 20 seconds in the Northridge section of the San Fernando Valley in Los Angeles, California. [1.9, 1.10.]

The earthquake had Richter magnitude of 6.7, with the same epicentral region that had been rocked during the 1971 San

Fernando earthquake. Fifty-seven people lost their lives in this disaster.

1.8.2   Adverse Observations

(a) Reinforced Concrete Bridges

- Older bridges with unusual geometries and large skews respond to earthquakes in complex ways that were not accounted for when designed.
- Retrofitting improves earthquake resistance.
- The significance of high vertical accelerations needs further investigation. (Fig. 1.4, Left.)

Fig. 1.4 (Left pair) 'Bird-caging' of top and bottom of R.C. columns; (Right) Reduced beam section in steel 'I' beams

(b) Steel Bridges

- Practice of only a few bays resisting lateral load risky.
- Standard detail of beam-column connections leads to severe inelasticity and stress concentrations.
- Welding practices leads to brittle failures.
- Yield zones under seismic action need considerable further investigation.

1.8.3   Recommendations and actions

The following list for this case study illustrates how forensic engineering can lead to better structural design and construction and to improved safety:

- More forensic engineering

- Further analytical and experimental study
- Consideration of combined horizontal and vertical loadings
- Changes and improvements to maintenance regimen
- Revisions of formulas and coefficients, and changes to codes of practice
- Improvements to detailing practice
- Modifications to construction practice
- New construction methodologies
- Avoidance of expansion joints
- Development of new designs (Fig. 1.4, Right.)

1.9 REFERENCES

1.1. Fowler, David W., *Forensic Engineering*, Powerpoint presentation at University of Texas.
http://www.matscieng.sunysb.edu/disaster/

1.2. N. Krishnamurthy, *Introduction to Risk Management*, ISBN 978-981-05-7924-1, 2007, 86p.

1.3. M. Matousek and Schneider, J., (1976) "Untersuchungen Zur Struktur des Zicherheitproblems bei Bauwerken", Institut für Baustatik und Konstruktion der ETH Zürich, Bericht No. 59, ETH.

1.4. Rachel Martin,
http://www.eng.uab.edu/cee/REU_NSF99/rachelwork.htm.

1.5. Henry Petroski, *Design Paradigms*, Cambridge University Press, 1994.

1.6. Henry Petroski, *To Engineer is Human*, St. Martin's Press, 1982.

1.7. Masayuki Nakao, "Chain Reaction Collapse of a High-rise Apartment due to a Gas Explosion May 16, 1968 in Ronan Point, East London, England", *Failure Knowledge Database/100 Selected Cases.*

1.8. Cynthia Pearson and Norbert Delatte, "Lessons from the Progressive Collapse of the Ronan Point Apartment Tower", *Proc. of 3rd ASCE Forensic Engineering Congress*, 19-21 October 2003, San Diego, California, p. 190-200.

1.9. Cooper, J. D., I.M. Friedland, et al, "The Northridge

Earthquake: Progress Made, Lessons Learned in Seismic-Resistant Bridge Design", *Public Roads On-Line*, Federal Highway Administration, U.S. Dept. of Transportation, Summer 1994, Vol. 58, No. 1.

1.10. Ron Shepherd, "Investigation of the Seismic Response of Welded Steel Moment Frames", *Proc. of 2nd ASCE Forensic Engineering Congress*, 21-23 May 2000, San Juan, Puerto Rico, p. 483-492.

―――――

† This is a reprint of the author's paper presented at the Structural Engineering World Congress (SEWC) organized by **SEWC Inc.**, at Bangalore, India, 4-9 Nov. 2007, reprinted with permission.

2

INVESTIGATIVE METHODS IN FORENSIC CIVIL ENGINEERING†

2.1 INTRODUCTION

The word *forensic* comes from the Latin *forēnsis*, meaning "of or before the forum." In Roman times, a criminal charge meant both accuser ('plaintiff') and accused ('defendant') presenting the case before a group of public individuals in the forum. This is the source of the two modern usages of the word *forensic* – as a form of legal evidence, and as a category of public presentation. In modern use, "forensics" or "forensic science" is effectively a synonym for "legal" or "related to courts".

Forensic engineering is the application of engineering principles to determine the causes of an accident, failure, or other performance problems. Generally, the purpose of a forensic engineering investigation is to locate causes of failure with a view to improve performance or life of a component, or to assist a court in determining the facts of an accident.

Civil Engineering may be considered to be the broadest of almost all engineering disciplines, covering the welfare and activities of huge masses of people all over the world. It also is the one industry which involves various disparate skills and trades, each having their own agenda and each vying for its rightful space and time. As a consequence, any mishap in civil engineering will generally affect a large number of people, and have far-reaching implications and long-lasting consequences.

Civil engineering often ends up with having the biggest failures leading to enormous loss of lives, injuries, and massive property damage in all nations. Forensic civil engineering is the logical investigation and legal presentation of various aspects of these accidents.

This paper will highlight the common techniques available for forensic engineering, with particular reference to civil engineering.

2.2 QUALIFICATIONS FOR A FORENSIC CIVIL ENGINEER

2.2.1 Forensic engineer

Who can be a forensic engineer? Columbia University in New York puts it as follows, [2.1]:

"An engineer's success in the field of forensic engineering is the result of the combination of many components in his or her background:

- First, a good education in engineering and its related subjects;
- Then years of hands-on experience in analysis, design, construction, testing, inspection, condition assessment, and trouble-shooting;
- Understanding of the design-construction process;
- Comprehension of legal implications;
- Good communication skills;

- A knack for problem solving;
- A positive attitude to team work;
- A strong sense of ethics;
- Self-confidence without arrogance;
- Confident and credible disposition; and,
- A high level of intellectual sophistication."

Carper [2.2] lists some more factors not in the preceding list:

- Detective skills
- Other skills such as familiarity with psychology and sociology, photography etc.
- Personality characteristics such as flexibility and objectivity, ability to face questioning under stress (as expert witness), and to work effectively with others.

To this I will add, a desire, almost a passion to share one's expertise to benefit a cause he believes in, be it to improve design, to affirm safety, to right a perceived wrong, and so on.

[NOTE: The male pronoun and references there from will include the female counterpart, except when context defines one or the other gender. – NK]

2.2.2   Forensic Civil Engineer

A career choice source [2.3] lists the following as job requirements to be a forensic civil engineer:

(a) Education

Bachelor's degree in civil engineering. A master's or a doctorate in some aspect of civil engineering – structural, geotechnical, or earthquake engineering – is often an advisable educational qualification in a competitive job marketplace. During these courses, he should become familiar with the way different types of materials behave under stress. The candidate should also have a strong mathematical ability, IT and communication skills. He will need to explain complicated technical matters to lawyers, legislators and the general public as part of his job.

(b) License

Candidate must obtain a Professional Engineer license in the state where he is employed. [

[Where such licensing does not exist, some extra qualification like Membership in national or international professional societies, publications in journals etc. may validate credibility. – NK]

(c) Site Experience

A forensic civil engineer's expertise is gained primarily from site inspection of structural damage, accidents, and disasters. He learns to recognize how and where materials and structures fail, where a building may have been incorrectly located, and the signs of possible criminal negligence or deliberate damage. This site experience provides him with the knowledge to analyze the causes of the damage.

Human communication skills and an empathetic personality are also very important as he must interview witnesses on the site who may be very distressed.

(d) Fitness

Physical fitness is vital for the forensic civil engineer. He may have to climb on roofs, move or carry heavy loads up to about 50 pounds ([18.7 kg) at a damage site, and even crawl through confined, damaged spaces such as air conditioning vents.

Site inspections have to be conducted in all weathers. Damage sites are dangerous, so he needs quick instincts to move very quickly to escape serious injury from any falling structures.

My journey to my current forensic involvement started in 1960s in USA when I acted as consultant to accident investigators – not yet known as 'forensic engineers'.

It was only by choice that I did not actually appear in court, reluctant to getting involved in legal tangles, especially as I had chosen not to become a citizen.

But by proxy, I learnt many of the tricks and traps of the forensic engineering trade. So, when the opportunity arose for accident investigation in Singapore, one might say I was willing, ready, and able.

2.3 CAUSES AND INVESTIGATIVE METHODS

Accidents are called thus because they are unpredictable, and usually the result of some unexpected combination of unusual circumstances, with generally one or more humans involved in its causation and/or in suffering its consequences.

Accidents must be investigated because the victim (and/or his relatives and friends) and the society at large must know how the accident happened, to apportion responsibility, and to ensure that the triggers and circumstances of the accident may be avoided or mitigated.

Most failures are quite straight forward, causing human, property, environment, or other harm or loss, and falling into one or more of a few categories of causes and requiring only a few well-known and simple investigative procedures, as in Table 2.1.

Table 2.1 Accident Causes and Investigative Methods

	Cause	Investigation
a.	General, for all causes	Review of all existing documents and actions Code conformity Conformity with good practice Accident re-creation, failure simulation Destructive/non-destructive tests Back analysis
b.	Wrong or under design	Design checks and back analysis
c.	Wrong erection or poor workmanship	Check procedures, supervision, etc. Check connections Check temporary supports
d.	Overloading	Recent and long-term history of use

	Cause	Investigation
e.	Wrong or bad materials	Material testing (Including physical and chemical properties; high-tech, e.g. gas chromatography, mass spectrography) Strength testing
f.	Vagaries of weather, natural disasters; Acts of God	Historical records versus design briefs and construction/fabrication records

2.4 DATA COLLECTION

Data is the backbone and lifeblood of any research or investigation. A case stands or falls, the guilty are punished and wrongs are righted, only with adequate quantities of right data. In accidents, data is the rare factor, because accidents happen without warning or control, and much data can get compromised or lost by essential accident response services and well-meaning and curiosity seeking public – or by carelessness in the chain of custody.

2.4.1 In a Perfect World

You may be lucky. You may be in a country which has a full-time forensic team on standby to rush to an accident site right behind or along with the ambulance and the police when the first information reaches the authorities. And you are lucky enough to reach the site fast.

Then your job is easy. The paramedics have first shot at the site, trying to save dying victims and patch up the injured to transport them to ICUs and clinics. Police cordon off the area so nobody tramples on accidentally – or modifies intentionally – any of the evidence.

You as part of the forensic team have almost equal powers with the police to approve or deny access to the site artefacts. Even the police are trained to respect your domain. They wear shoe covers and hand gloves and watch where they walk and what they touch. They sequester potential witnesses for your questioning as soon as they have elicited as much case evidence as they can from them.

You go in with your cameras and specimen kits, special lights and field test equipment, and record everything you are likely to need.

Another member of your team handles the witnesses and debriefs them. Or vice versa.

Some specific individual or group is already in charge of the most important immediate tasks after an accident:

- Rescue operations,
- Medical treatment of the injured,
- Prevention of further injuries,
- Securing the site to protect the evidence from wilful or unwitting compromise.

You collect the following data, as appropriate:

- Status of the accident site:
  o Documentation of what, where, and how things and people are, by means of photography, videography, and audio recordings;
  o Collection and safeguarding of documents and trace evidence;
  o Material Safety Data Sheets (MSDS), equipment manuals, etc.;
- Data on personnel present
- Immediate past history:
  o Collection and safeguarding of still and CCTV and other video camera tapes or electronic media;
  o Data on deceased and injured;
  o Temperature, sound, pressure and other records for as long before the event as are available;
  o Eye/ear witness's evidence in recorded interviews;
  o Details of personnel and organisations involved.
- Long-term history:

33

- o Original designs, subsequent modifications;
- o Construction records, safe work procedures (SWPs);
- o Material indents and deliveries, test records;
- o Use histories, and maintenance and repair records;
- o Risk assessment and management records;
- o Personnel and organisations in various stages of construction and use.

It may fall to your lot, as a civil engineer, to stabilise the site against repetition of the mishap or further progressive collapse. If you see a beam precariously sagging, you would have to try to prop it up from further sagging – trying to push it up to its original position may neither be possible nor wise, because it may produce further damage.

Advanced countries have strict standard procedures for securing the site for the forensic team. Until this team releases the site, nobody can enter it without specific approval, and even then, only with proper accessories to prevent site contamination, and under strict supervision.

Without such control, forensic engineering would be a sham.

You should be that lucky!

2.4.2 Welcome to the Real World

More often than not, the accident is past history.

Except in rare instances – admittedly increasing due to the ubiquitous CCTVs watching over almost every aspect of our lives – in most countries, the accident is long gone and the site cleaned up before any real investigation gets under way. This puts an extra burden on the forensic engineer, and he needs to have access to a variety of methods and techniques to seek out the truth from the available data after the event.

However, even though the accident is over and clean-up has begun by the time you are called in as an expert witness, generally it would be worth your while to visit the site to view the lay of the land, the scale of operations, etc.

In a case concerning a formwork collapse a few years ago, I was called in a few months after the accident, by which time the dead bodies had been removed and the blood had been hosed off, and the collapsed formwork had been removed, replaced by a spanking clean and perfect replacement formwork, and the casting of the permanent structure already started.

While I was testifying as expert witness, the other side lawyer asked me if I had visited the accident site, I simply said *"No."*.

When the lawyer wanted to make an issue of it, saying that I had failed in his duty as an accident investigator, I explained that it would have been a waste of time, and that I had enough certified photographs, videos, and eye-witness accounts to extract all the information I wanted for my forensic analysis.

In most developing and under-developed countries when an accident happens, it is a free show for curiosity seekers (and let us admit it, even to most of us, intelligent, mature adults!) and chaos reigns almost seconds after the accident. It is a free-for-all, with anybody and everybody standing around busily watching the show, most chipping in with well-intentioned but unhelpful suggestions, and many actually entering the accident site and handling artefacts and people with avowed intention of saving persons in distress or articles liable to damage.

But unfortunately, they are also irrevocably destroying valuable evidence that would have helped discover how the accident happened and possibly guide how to prevent future accidents.

Often, it is also the occasion for unscrupulous persons to steal things – even off of the dead and wounded – and for supervisors and managers to modify the scene to protect themselves from litigation. Even in the best of countries, watches, hand-phones and wallets vanish. It is not also unusual for an investigator to find that a safety harness has been placed next to or even <u>on</u> a dead fallen worker, or a brace in a scaffold which had not been in the photograph taken yesterday, to be suddenly in place today!

In such countries forensic engineering is still at best an academic topic for journal publications and conference presentations, but not a practical solution to improve safety.

Also, forensic investigators are not too popular because they probe into tragedies, scratching emotional wounds, and asking inconvenient and embarrassing questions. Not having the authority of the police or the power of medical personnel, forensic investigators are often the last, unwelcome, guests in the chain of command.

Still, you do the best you can, and collect whatever data you can in the best way possible. Do keep records of the times and dates of how and where you got the data, so that you won't be held responsible for ineffective investigation later on. And once you collect any data, have a way of authenticate it, with more than just your say-so, because your word without corroborating confirmation will not be worth much in court as evidence.

2.5 CODE AND GOOD-PRACTICE CHECKS

In any civil engineering investigation, checks on Code compliance or good practice would be the first line of action to take. Many failures happen because the designer or contractor has not strictly or correctly followed the applicable Codes, Standards or Practices, or the contractor has not followed the instructions of the designer.

Such violations are normally sufficient to indict the wrong-doer – although courts may want the plaintiff to prove negligence or intent.

In the infamous Kansas City Hyatt Regency walkway failure (which will be discussed later in another context), the originally designed walkways were barely capable of holding up the expected load, and would not have met the requirements of the Kansas City Building Code, [2.4]. Further meddling simply aggravated the situation into a tragedy.

Conversely, the findings that wrong-doers turn out not be punishable under current Codes are often the trigger for Codes to be changed. For instance, as a result of the World Trade Center

Investigation (2001), a total of 40 code changes consistent with recommendations were adopted in the 2009 and the 2012 editions of International Building and Fire Codes, [2.5].

There are a large number of cases where Code violation leading to failure has happened. Some of them will be briefly presented.

2.5.1   Rana Building Plaza, Bangladesh, 2013

On 24 April 2013, an eight-story commercial building, Rana Plaza, housing five garment factories collapsed in a suburb of Dhaka, capital of Bangladesh. More than 1100 workers died and about 2500 were injured. [2.6] (Figure 2.1.)

Fig. 2.1 Rana Plaza Building Collapse

The head of the Bangladesh Fire Service and Civil Defence said that the upper four floors had been built without a permit. Rana Plaza's architect said the building was planned for shops and offices – but not factories.

Other architects stressed the risks involved in placing factories inside a building designed only for shops and offices, noting the structure was potentially not strong enough to bear the weight and vibration of heavy machinery.

More significant than the violations was the discovery that there were not adequate codes and regulations for many of the good practice violations that were encountered. Hopefully, this disaster will pave the way for better standards and more stringent enforcement, not only in Bangladesh but in neighbouring countries where similar conditions of work may exist.

2.5.2 Fall of Worker from Mobile Tower

I had occasion to investigate the fall of a worker from a mobile scaffold which collapsed with the worker on top, allegedly while the tower was being moved, which was against local regulations. The worker had a head injury which necessitated the replacement of part of his scalp with a synthetic shield, depriving him completely of his livelihood.

The employer offered a token payment. While it would have had a good multiplicative effect due to the exchange rate between the local currency and the worker's home currency, it was pitifully low by local standards, and the worker's lawyers requested my assistance.

What is relevant here were the charges by the employer that (Figure 2.2):

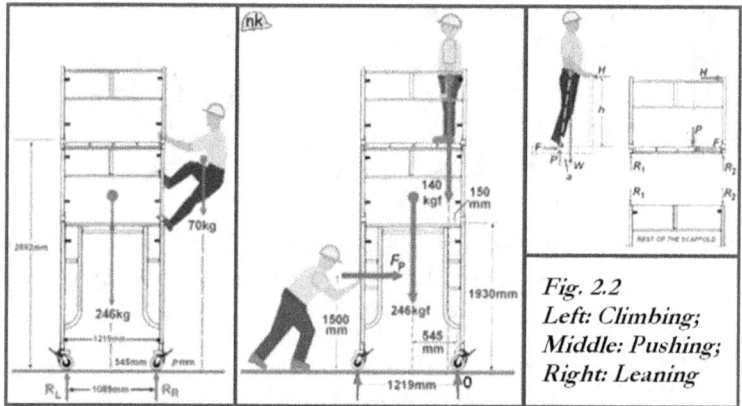

Fig. 2.2
Left: Climbing;
Middle: Pushing;
Right: Leaning

(a) The employee climbed the side of the scaffold to reach the platform,
(b) He stayed on the top while another worker moved it to a new location, and
(c) He leaned on the guard-rail during his work.

All three were unsafe acts and the company was making the payment more as a donation than as compensation.

True, all three acts violated local Codes; and if these charges had been proved, the worker would have had not only no compensation, but also to pay a fine and/or undergo a jail term.

After examining the photographs and witness testimonies, I accepted the case because I believed that it was not the worker but the employer who was at fault.

I should say I had an easy time shooting the company argument down because all three of the charges were traceable to the employer's violation of applicable codes:

(a) Worker climbed the scaffold only because the tower did not have the Code requirement of safe access by ladder or steps, as could be proven by photographs and testimony;

(b) Worker stayed on top, only because he did not want to risk injury by climbing up and down the scaffold side (*vide* previous violation); and,

(c) Simply leaning on, or slightly over, the guard-rail in the normal approved course of work, namely painting the ceiling, would not have resulted in his toppling over if the guard-rail had been the Code requirement of 1 m minimum height – it was only about 900 mm.

To prove the above, my first year engineering Statics was enough – although I was surprised at a similar analysis by their expert who had managed to prove quite the contrary, by some convoluted arguments which could not be substantiated.

It was relatively easy to prove that the worker climbing the side, or one worker pushing the scaffold while the other was still on it, or a worker leaning on the guard-rail, would not have toppled the scaffold as claimed.

But the employer would not simply accept my word that if the guard-rail had been the required 1 m tall, the worker would not have fallen for simply leaning on it and painting the ceiling.

He demanded that I prove any guard-rail shorter than the specified height could result in the worker falling over the rail if he leaned over it in the course of his assigned task. To achieve this, I developed a two segment model of the human body as shown in Fig. 2.3, left.

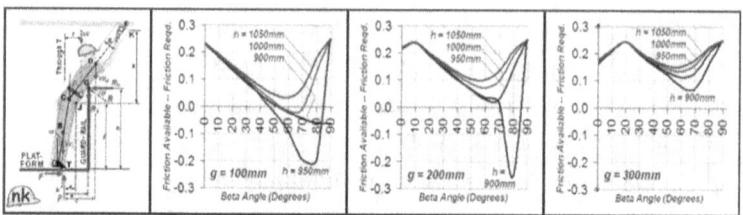

Fig. 2.3 Painter fall from scaffold – Angle, girth and guard-rail height effects

After carrying a large number of parametric studies on this model with various heights and girths of workers for different heights of guard-rail, using anthropometry (measurements of the human body) and biomechanics (static and dynamic behaviour of living systems), I was able to draw a number of general conclusions.

The charts in Fig. 2.3 clearly show that for a person with girth between 100 and 200 mm, only 1m tall guard-rail would be safe for any angle of leaning over. For our worker of medium girth (200 mm), 950 mm would be minimum. The 900 mm guard-rail was just not tall enough.

Faced with my report listing these (and a few other) Code violations and technical findings, the employer settled out of court for a much higher compensation, with which the worker was able to return home, and open a shop as a means of a reasonable livelihood for the rest of his life.

If the case had gone to court, for the unsafe conditions which violated the Code and resulted in the worker's fall, the employer would have been heavily penalised.

At the same time, the worker would also have been punished for his unsafe acts, because in theory, a worker is not supposed to do an unsafe act but refuse to carry out instructions under those conditions – courts may not get emotional at the nervousness and language problems of the worker in this regard which prevented this worker from refusing to work.

This research into falling behaviour was a bonus to me. The entire investigation including my research on the fall behaviour has been published in a journal, [2.7].

WARNING: It must be remembered by forensic engineers that each case is unique and must be analysed in the context of circumstances prevailing at the place and at the time the accident happened, and under the regulations governing design, construction, and use in that situation.

For instance, in a situation by Zallen [2.8] very similar to the one I have described above, two workers were working on a mobile scaffold, and one of them was severely injured when the scaffold fell while exiting.

They climbed up a step ladder to reach the scaffold, and normally exited the same way. However when one of them swung his body around a guard-rail post and exited under the end frame rail, the scaffold toppled.

Figure 2.4 shows two scenarios of the fall. The left figure shows the worker exiting from the front ('end') and swinging around the left post, thus imposing a horizontal dynamic force on it, which toppled the scaffold.

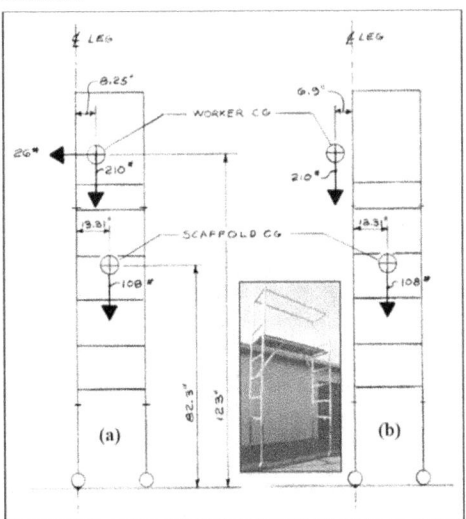

Fig. 2.4 Unsafe act on mobile scaffold

The right figure shows him ducking under the side rail and simply climbing down; even this was unsafe because the overturning moment by the worker was more than stabilising moment from the much lighter Aluminium frame.

Both these acts were shown to cause overturning moments in excess of the stabilising moments. As in the USA scaffolds are required to be designed for such acts by workers, the employer was found culpable – he should have provided out-riggers extending the base to allow for the overturning moments.

The differences between our two cases were two-fold:

(1) Unlike in my case where the scaffold was steel, this scaffold was of Aluminium, hence much lighter than the worker's weight.

(2) In my case the law forbade climbing on the side of a scaffold, but in USA, there is no law against such climbing.

2.6 WRONG DESIGN

One would think that in this day and age of advanced technology designers would not make a mistake in engineering basics. While most designs pass through more than one set of hands and eyes, occasionally, either the complexity of the problem throws an engineer or even a team off track, or at the other extreme the problem appears so simple that nobody pays any attention to the solution.

A borderline situation is when something wrong has been working without problems for long, and you get so used to it, you do not see the mistake staring at you, until something bad happens and a fresh set of eyes jumps at the mistake instantly.

2.6.1 Hotel New World, Singapore, 1986

On March 15, 1986, the six-story Hotel New World in Little India, Singapore, collapsed due to a design error. [2.9] (Figure 2.5.)

The structural engineer had forgotten to add the building's dead load to his calculations when determining how strong he needed to make the support pillars that held up the building during construction in 1971. On top of it, the owner added quite heavy water tanks and air-conditioning units on the roof, and a bank tenant added a heavy vault, both without checking or approval, [2.10]. The building was a steel-reinforced concrete design.

Fig. 2.5 Hotel new World, before and after collapse

Thirty-three people were killed and 17 others were injured.

In regard to the bank's liability, the Court of Appeals held that:

"[T]he collapse of Hotel New World in 1986 must be looked at with a 1986 pair of spectacles. Having found that until 1986 there was no instance of a collapse of a building such as the hotel which stood for more than 10 years, the bank could not be imputed with knowledge of the various unusual dangers that were raised in that case, such as tremors, vibrations and cracks in the building."

Following this disaster, many lessons were learnt. Buildings built in the 1970s were checked for structural faults, and some of them were declared structurally unsound and had to be evacuated. The government also introduced tighter regulations on building construction; since 1989, all structural designs are required to be counter-checked by Accredited Checkers.

2.6.2   Nicoll Highway Collapse, 2004

The 20 April 2004 Nicoll Highway collapse in the C824 project for the cut-and-cover underground MRT Circle Line in Singapore had two critical design errors according to the International Committee of Enquiry, [2.11]. (Figure 2.6.)

These were:

(a) Under-design of the diaphragm wall using an inappropriate method in the computer program; and,

(b) Under-design of strut-waler connection in the strutting system (Arrows, Fig. 2.6.)

Fig. 2.6 Nicoll Highway Collapse, Left – General View, Part Plan, Right – Buckled Walers

These design errors resulted in the failure of the 9th level strut-waler connections together with the inability of the overall temporary retaining wall system to resist the redistributed loads as the 9th level strutting failed. The catastrophic collapse then ensued. Three workers and one official died and three others were injured.

The sad fact was that the high deflections of the diaphragm walls had given ample warnings of impending disaster for weeks, which were ignored.

The Executive Summary of the report also charged the designers with *"Abuse of the back analyses in Type M3 where the collapse took place."* Back analysis is a reverse procedure, to solve external load or partial material parameters, based on known deformation and stresses at limited points and partially known material parameters. It is used in geo-technical engineering. With sufficient redundant data it has also been also applied to structures.

The company was fined S$200,000, and three senior executives were fined S$120,000, S$160,000 and S$160,000. Government officials, who were also found guilty of negligence in their duties could not be fined, but were warned and/or counselled.

A legal criterion which could greatly impact criminal liability to such negligence has been articulated many times: We should judge the conduct of designers by the standards prevailing and accepted at the time when it was designed (in this case the Factories Act), and not by any later standards. So, perhaps the penalties were not as severe as they would be today.

But out of this disaster came the development and

implementation of the Workplace Safety and Health Act in 2006.

2.7 CONSTRUCTION AND ERECTION DEFICIENCIES

Often, the design is fine, but the contractor finds he cannot implement parts of it. Technically, the contractor should bounce the ball back to the designer explaining his difficulty so that the designer can either revise his design or advise a practical method of achieving his ends.

The tendency of contractors though is to substitute the impossible design with a feasible alternative, with or without a rigorous analysis by their own competent person, particularly on how the substitute design component fits into the overall design which the original designer had submitted. Many accidents are on record due to this non- or mis-communication between designer and contractor.

Things are changing. The high costs of accidents have forced formerly separate entities to share mutual concerns, before tragedy hits. Many standards have adopted the 'Design for Safety' and 'Life Cycle Design' approaches whereby the designer is expected at least to highlight hazards and difficulties in the implementation of the design details.

A few case studies will be summarized below.

2.7.1 Skyline Plaza Apartments, USA, 1973

Around 2.30 pm on March 2, 1973, the Skyline Plaza 26-story apartment building in Bailey's Crossroads, Virginia collapsed while under construction, at the rate of one floor per week, except that recently, the pace had been accelerated, so that when the concrete was being placed for the 24th floor, the 23rd floor slab was only four days old instead of seven days, [2.12]. This was against the local Code. (Figure 2.7.)

The collapse killed fourteen workers and injured 34.

Fig. 2.7 Skyline Plaza

It was found that while full shoring remained on the 23rd and 24th stories, it had nearly all been removed from 22nd story in about 3/4ths of area. Thus relieved of its previous loads, deflection of the 22nd story would have decreased in the slab, causing the re-shores on the 21st floor to fall out.

Construction did not adhere to the engineer's stated requirements: *"Slab being poured to be shored for two floors and back-propped at center of span each way and at center of bay on next floor down,"* or the architect's specifications requiring *"in all cases, two floors shall be fully shored".*

Probable cause of the collapse was a punching shear failure of the 23rd floor, caused by the premature removal of forms supporting the 23rd story slab when the shear stresses exceeded the concrete capacity at the time of the incident. The accumulation and impact of falling debris from the collapsing 23rd and 24th floors overloaded the 22nd floor slab and induced the progressive collapse of successive floors down to the ground.

Average air temperature was 6-7°C, at which, concrete takes twice as long as in a laboratory to gain strength. Field-cured concrete cylinders should have been used rather than laboratory cured cylinders to ensure that the concrete had achieved sufficient strength before removing shores and forms.

The concrete subcontractor paid less than $20,000 in fines, despite the fact that the collapse had caused $8 to $10 million in damages, in addition to the deaths and injuries.

The architect and engineer, sued by one injured worker, were held responsible for the collapse, even though their explicit specifications for required shoring were not implemented at the site, and had to pay his claim of $250,000 – ironically, this last happened to violate the requirement that the designers had a responsibility to inspect the work and warn those involved of any unsafe conditions.

Today things would be much different – except that 'design for safety' and 'life-cycle design' are recent trends imposing more responsibility for construction and use on the designer/architect!

2.7.2   The Chicago City Post Office

On November 3, 1993, a portion of Chicago's new main post office under erection, collapsed, killing two ironworkers and injuring five others.

At the time, workers were laying beams in place before fastening them.

One of the insecure beams caused the collapse by creating a chain reaction and pulling down 60 to 70 other already erected steel members. The beams that collapsed were 32 ft (9.75 m) long and weighed four and a half tons (45 kN), [2.13]. (Figure 2.8.)

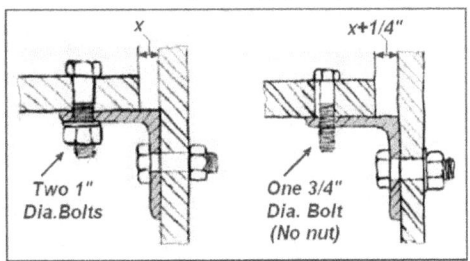

Fig. 2.8 Left – As designed, and
Right – As erected

Root cause was traced to the changes made by the fabricator in the erection procedure, solely for ease of erection, resulting in the beams being placed 1/4 inch (6 mm) further away from columns than required.

This change made the use of the 1 inch diameter bolts impossible. Instead, workers had to use 3/4 inch bolts to secure the beams. The change in hardware led to a series of weak connections, some without nuts to secure bolts.

The location where collapse began, unsurprisingly, was a connection where a nut was not used.

All this havoc for just a quarter-inch shift of a temporary support, and a quarter-inch thinner bolt!

2.7.3 Hartford Civic Centre Arena Roof Collapse

Another classical case where the erector modified the designer's connection and caused structural collapse was the Civic Center Arena in Hartford, Connecticut, USA, completed in January 1973, designed for first time by a 3D truss analysis computer programme, [2.14].

As shown in Fig. 2.9, the 300 ft. by 360 ft. (91 m by 110 m) roof space frame consisted of pods in 30 ft. by 30 ft. (9.1 m by 9.1 m) grids, 21 ft. (6.4 m) apart.

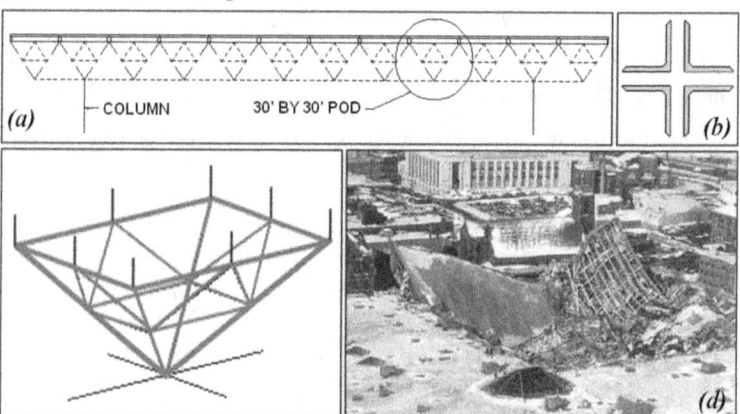

Fig. 2.9 Hartford Civic Center Arena, Sections and Collapse

They were constructed in the shape of a cross – which unfortunately is a most inefficient shape for bending and buckling.

On January 18, 1978, the largest snowstorm of its five-year life hit the arena. Early morning, the center of the arena's roof crashed down 83 feet to the floor of the arena, throwing the corners up into the air. Luckily the arena was empty.

Cause of the collapse was traced to relatively minor changes in the connections between steel components, the fabrication deviating from design. The most frightening result of the changes was in a particular connection in which a few centimetres shift of the fabricated connection cut down axial force capacity to less than a tenth of the design value. (Table 2.2.)

Table 2.2 Comparison of (a) As designed, and (b) As-built

	Connection A	Connection B	Connection C	Connection D
(a) Original	Allowable moment: 0 Allowable force: 160,000-lb	Allowable force: 185,000-lb	Allowable force: 625,000-lb	Allowable force: 565,000-lb
(b) As- built	Allowable moment: 9,490 lb-ft Allowable force: 15,440-lb	Allowable force: 59,000-lb	Allowable force: 363,000-lb	Allowable force: 565,000-lb

The designers should have checked whether their proposals could be translated into practice. Alternatively, the fabricator should have had the designers check and approve the changes he was proposing for practical reasons.

As in many other cases, the structure itself gave ample warning of impending failure by excessive deflection, which were ignored.

Designers too were over-confident about the computer programme which at that time had not considered all buckling modes, and ignored the large deflections during erection.

The collapse triggered a complete nation-wide review of similar structures and computer applications, and resulted in upgrading the safety aspects in design.

2.8 COMBINED DESIGN AND CONSTRUCTION DEFICIENCIES

Often in accidents it is found that flaws in design and construction combine to result in collapse although separately they might not have been that much of a problem. Examples of these causes abound in the forensic literature.

2.8.1 Hyatt-Regency Walkway Collapse

A classical example of a combination of design and construction error is the Hyatt Regency Walkway failure at Kansas City, Missouri, USA, on 17 July 1981, which killed 114 and injured more than 200, [2.15]. (Figure 2.10, left two parts.)

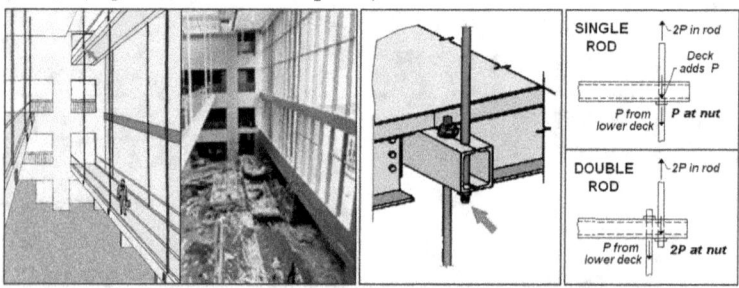

Fig. 2.10 Left Pair – Hyatt Regency Walkway, Before and after Collapse; Middle – Nut at Upper Walkway; Right – Static Analysis of Single vs. Double Hanger Rods.

On that fateful evening, during a 'tea dance', at 7.05 pm, the fourth floor walkway of the 4-storey high atrium in the 40-storey Hotel collapsed with excited foot-tapping couples, on to the second floor walkway, dragging it and its occupants to the bottom dance floors, leaving 114 dead and more than 200 injured. (Figure 2.10, left two parts.)

The hotel was opened in July 1980, with three walkways suspended from the ceiling by hanger rods, one set of rods on one side of the atrium for floors 2 and 4, and a second set on the other side for floor 3, as shown in Fig. 2.10 (Left). Both floors 2 and 4 were to hang from a single rod, with each floor load being supported by a washer and nut under the respective floor, the location being indicated by arrows in the first and third parts of Fig. 2.10.

But the contractors felt that to place a nut in the middle of the single rod to support the fourth floor was too much trouble, and proposed two rods, one going from the ceiling to the fourth floor and the other going from the fourth to the second floor, as shown in the third part of the figure.

Whether the contractors got approval for the change from the designers was not quite clear, but the contractors went ahead and implemented their change.

The problem was that while the single rod, inconvenient to erect though it was, would have been loaded with only a single floor load at each of the second and fourth levels, the two-rod solution, while simpler, dumped the second floor load also on to the same washer and nut holding up the fourth floor, in effect, doubling the design load, which became the direct cause of the collapse during the dance.

Although this could be referred to as a design flaw in that nobody checked the design capacity of the revised design, it was triggered more by the contractors wanting (and rushing to) changed the design without proper re-analysis.

The error that killed 114 people would not have occurred if the two-rod problem had been given to a first-year engineering statics student as a home assignment! So simple that veteran engineers did not think to check! Moral of the story: Nothing is too simple to check!

2.8.2   Rebar-support Collapse

(a) The problem and my role in it:

[I have omitted details of the case out of consideration for my fellow-engineers and academics involved, who are still in active service. – NK]

I   was   invited   by   the   authorities   to   investigate   for   the prosecution, the collapse of a rebar support system used for casting 3 m and 5 m R.C. base slabs, which killed two workers and injured 29 others. I investigated the accident and discovered a number of design and erection deficiencies.

The structure had problems from:

(i)   Inherent asymmetry of configuration, leading to instability;

(ii) Design errors, including use of a wrong coefficient in a standard formula; and,

(iii) Inadequate design procedure, in particular for the cross-bracing system to prevent buckling and sway.

As my testimony progressed, the judge sought from me a few additional analyses regarding scenarios of alternative failure scenarios – which I provided gladly despite having to miss some sleep two nights in a row for the computations.

It would be wrong (and unwise) to think that such extra requests (more like commands when coming from a judge) are an indication of your high stature or that you are doing a favour to the court. Far from it.

It may simply mean that while examining an expert they are claiming their right for more information – or even that you were not clear enough or complete enough in the first place.

In fact, the more you say, the deeper you may be getting into trouble! So, it was no feather in my cap that I was able to come up with useful results at short order. It would have been egg on my face and a real tragedy if I had not come up with the answers.

(b) Disposition of Case:

As expert witness, I testified to all the above and other deficiencies in design and construction. However, there were other erection considerations such as placement of heavy loads on the cage, on which there was no definitive information, but which might have led to the collapse. In the first portion of the case, the main contractor pleaded guilty to *"failing to ensure that the worksite was properly built and safely maintained"*, and was fined.

None of my preceding technical testimony was contradicted.

However, some of my recommendations were for future improvement, and violation of those norms could not be held against the management personnel on the convention that a case must be judged according to the laws and practice prevailing at the time and place the accident happened. The rest of the charges were dismissed on certain legal inconsistencies, and hence the details of design and construction deficiencies never became a cause for individual penalties.

2.9 ACCIDENT RE-CREATION

It would be wonderful if we can simply rewind life's tape a few minutes and review how exactly an accident happened. Actually, the chances of having a record of accidents are increasing because more and more CCTV cameras are being deployed. Also, with the ever-increasing versatility and power of hand-phones, a few or many might record the accident 'accidentally' while they were shooting something else, or purposely because they just happened to be there with a camera on stand-by mode.

In the West, there is money to be made from such records, selling them to the media, or to one side or the other (or both!), in the subsequent investigation – East may not be slow to catch up!

Barring such direct records, the next best thing is a re-creation (surely not 'recreation'!) of the accident – also called 're-enactment' or 'reconstruction'. This can get pretty complicated.

For one thing, things are never the same at a microscopic or instantaneous level. Most materials may be reproducible, but the exact sequence and intensity of actions may not be. For another, some acts are not feasible to reproduce, such as a live person falling off of from height. One cardinal rule is that no human, and these days, no animals either, must be harmed in the process of re-creation.

Still, a lot of information can be gleaned from an intelligent re-enactment of the accident.

Here, scale becomes a problem. For instance, cell-phone and

auto makers, even aircraft manufacturers, will be happy to sacrifice one or more of their products to undergo various destructive tests. But a civil engineer, involved with huge projects, generally does not have the luxury. He can only put a small segment of the structure through the accident scenario.

In citing examples of this, I would like to start with a non-civil engineering example of how sometimes the simplest and the cheapest of tests can prove a major truth.

2.9.1   NASA Challenger Disaster

During the heyday of American space supremacy, sending crews into space orbit for various lengths of time to carry out experiments and bringing them back to earth became a routine chore like sending kids to school and receiving them back in the evening. The Space Shuttle program became a bus ride.

It was during the tenth mission, launched in very cold weather against the warnings of at least some of the engineers involved, at about 73 seconds after launch, the shuttle blew up, and the pieces fell into the ocean, killing all the seven crew members. (Figure 2.11.)

The culprit was a gasket called the 'O' ring, which lost its elasticity in the extreme cold and failed to seal the hot combustion gases from cabin crew and instrumentation. The web is replete with details of the disaster, starting with a brief account from Wikipedia, [2.16].

The anecdote I wish to relate here is the way the gasket's dangerous behaviour in cold temperature was demonstrated by Nobel Prize winner Richard Feynman, while giving expert evidence on NASA's Challenger failure, [2.17].

To prove how easy it was to accept this, Feynman asked for a glass of ice-water while waiting for his testimony for the Presidential Commission, and soaked the sample of the rubber of the 'O-ring' which he had picked up on the way from a hardware store. When his turn came to speck, he simply pulled the hardened sample out and showed that it had lost its elasticity, at least temporarily, destroying the integrity of the seal, [2.18]!

Fig. 2.11 Left: Challenger Launch, crew, and blow-up. Top: Feynman shows cold effect on gasket.

If this sounds like something out of the American show CSI or Perry Mason stories by Erle Stanley Gardner, I can confirm that such is possible in the USA. Even in other countries, as long as you understand and abide by the local culture, you can present your re-enactment results with telling effect.

2.9.2   Field Test for Support Rotational Capacity

It may be sacrilege to speak about my Mickey Mouse experiment in the same breath as Feynman's demo above, but it will illustrate the point that sometimes a simple field test may prove an important valid point.

In the rebar support system collapse case mentioned earlier, one of my contentions to explain the collapse was that the vertical supports had very little lateral resistance. The defence had conducted lab tests to document its tension, compression, and shear tests, but had overlooked the rotational capacity. I did not have the time (or the means!) to conduct a similar lab test for it, so I chose to set up my own 'quick-and-dirty' field test.

Fig. 2.12 Author's field test for rotational capacity

I had my lawyers arrange with a local contractor to set up a single bar of the same size as the collapsed bar with same bottom support conditions. I got a worker pull gradually at top with a rope, with the force of pull being measured by a simple spring balance. (Figure 2.12.)

When I presented this in court, the defence lawyers argued at my conducting such an unprofessional test which would prove nothing of value in the case. Briefed by their expert, they asked if I had calibrated the spring balance, which was a common 25 kg spring balance bought from a local store for household use.

I simply told them I had not wanted to conduct a high-tech lab test to fractions of a kg force, and was willing to accept the usual precision of say plus or minus 1 kg for the household spring, which would be sufficient for my purpose.

After all, the bar fell at about 15 kg pull, which was a small fraction of what would have been necessary to prevent collapse of rebar support system under lateral force. What difference would it have made if it had been 16 kg, or even 18 kg? It confirmed my guess that the rotational mode of failure was much easier than any other mode, and that was the way the system had failed.

The defence had also conducted a re-creation of the accident in a university lab taking a sizable segment of the support and top re-bar system, at considerably greater time and expense. On that, I presented my own arguments as to how a too-well organised and planned lab test might not explain the collapse due to site deficiencies – but that is another story.

None of the arguments from both sides was really conclusive, because there were various other factors at work contributing to the accident. But author had his two-cents' worth at much less expense than the other side!

2.9.3   Re-creation of Ladder Accident

When someone gets hurt while using a commercial device, quite often (particularly in the West), product liability claims can get into the millions, and then, experts would have an awesome job to prove 'beyond reasonable doubt', the culpability of the manufacturer or facilitator of the device.

One case which illustrates such a re-creation is the forensic analysis of a fall from a ladder in UK in the 1990s. A 73 kg house owner cleaning the second floor window of his house from his recently bought extensible ladder, fell while climbing. He approached the Legal Aid Board for an opinion on whether he can sue the ladder company for a faulty product.

One of the methods the Board's expert used was to re-create the fall using a weight suspended from different rungs of the ladder corresponding to different positions of the user, and also for different angles of the ladder with the horizontal. (Figure 2.13.)

It was established from the owner's description of how he had set up the ladder that the owner had started with an angle of ladder about 56° with the horizontal as against the approximately 75° (corresponding to 1 horizontal to 4 vertical slope) recommended for safe use by the manufacturer – and most others the world over.

Tests with the dummy weight demonstrated that when the weight was hung near the top, as the owner was positioned when he fell, at the 56° angle of the ladder, it became unstable and the foot slipped away from the wall, the top sliding down the wall – in the right portion of Fig. 2.13, the ladder is static only because the top has been stopped by a ledge on the wall.

Fig. 2.13 Re-creation of ladder fall
(Far ladder is for control.)

As the owner had not followed the manufacturer's instructions for slope, he would not be able to sue them, according to the Legal Aid Board. Had it been otherwise, in such product liability cases, the payout tends to be huge (often in the millions!) in the Western countries.

Rest of the Forensic Analysis:

A side-note to this example is worth adding. The Board investigation conducted so many other analyses and tests as follows, with reference to Fig. 2.14:

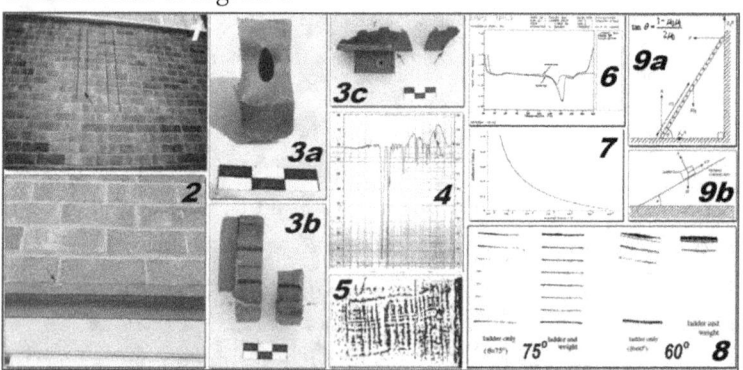

Fig. 2.14 Various tests and analyses run in ladder investigation

1. Ladder draw down marks on the wall,
2. Impact damage to the sill above patio door,
3. (a),(b) – Ladder tip fracture surfaces, (c) – Failed ladder tip, abrasion from brick,

4. Infrared spectrum of ladder tip material,
5. Uncontrolled side slip marks on wall (on white board tacked for that purpose),
6. Melting curve of Differential Scanning Calorimetry (DSC) for tip material,
7. Coefficient of friction vs. Load, for ladder tip material,
8. Foot-prints of ladder feet at 75° and 60°, and,
9. (a), (b) – Various statics analyses of ladder friction on surfaces of different slopes.

This is an unusually comprehensive list of forensic investigations for a ladder accident – but it shows how much a serious investigation would involve!

2.10 ACCIDENT SIMULATION

While accident re-creation is a physical prototype of model of the accident artefacts and events, accident simulation is a virtual reconstruction of the accident with a computer model or a physical small-scale model (or occasionally a large-scale version when the object itself is very small and you cannot see much in an actual scale re-creation). This is also a godsend where physical re-creation would be too large, too expensive, too time-consuming and/or too dangerous to people or property or environment.

2.10.1 Computer simulation of accidents

Today, most major accidents are simulated with computer models. The computer is used not only for very highly complicated mathematical analysis of accident parameters, but through computer graphics and animation, also for quite realistic visuals in still and video format.

Apart from impressing all viewers, the computer formulation also allows the analyst to 'play around' with the objects and their movements with very little effort beyond the first modelling, so that by repeated trials, one or a few scenarios which would have led to the accident can be quite reliably determined.

The computer applications will be discussed in detail in another paper, [2.19].

2.10.2   Model Analysis of Complex Structures

Another alternative is some kind of model analysis in which a scaled version of the situation is subjected to the same event as in the accident.

Before computers became so common, so powerful and yet so inexpensive, model analysis flourished both as a science and as a practical procedure to solve complex engineering problems. Even now, except that people would not know or care for the power of model analysis, it can serve as an effective alternative to complex computer analysis, which even today depends much on the proper use of a good software.

While doing my PhD in the USA, a Professor took our class to visit the well-known structural designer Ken R. White's office in Denver.

Mr. White showed us some small-scale models of innovative structures he had built out of tooth-picks and drinking straws, and then loaded them and pushed them around until they failed, to get an idea of their failure modes and loadings.

This was before the days of finite elements and 3D computer graphics, but the idea was the same: You have no record of the accident itself.

You develop a model, and experiment with various scenarios, until the failure is similar to the actual accident.

Even today this is the most common procedure. In many cases however, the creation and testing of the model, physical or digital, can get quite expensive.

For physical models in pre-computer days, scaling effects and material properties were handled by a now-defunct science called 'Similitude'.

I developed a lab-oriented course on 'Similitude in Engineering' during my tenure at Auburn University in Alabama, USA (1967-1975), and employed the method to good effect in my research and to guide student theses.

After the advent of computers, by the 1980s, we got the power to model huge structures in great detail and analyse them by computers, and then model analysis fell into disuse.

2.11 SUMMARY OF AN ACCIDENT INVESTIGATION AND REPORT

One of the first accident investigations I took up was that of a construction worker who fell seven storeys while working at a demolition of a multi-storey building. The worker was paralysed, and the company was directed to get their safety management system checked.

Many causes for accidents may be classified as unsafe conditions by the management and unsafe acts by the workers. In this case there were enough violations of both types.

The managements of the scaffolding and demolition contractor involved were very experienced, so experienced in fact that they began taking certain short cuts and worker behaviour for granted.

The injured worker had been with them for many years, and he had chosen to do things in his own way, and his supervisor had relaxed site discipline *"because he was such a good worker"*.

He had been given a double lanyard (which consisted of two short leashes from his waist-belt to clamps at least one of which had to be attached to a dependable anchor while he was working). This was considered "100% tie-off", a sure-fire fall prevention method. (Figure 2.15.)

On that fateful day, allegedly he had used only one of the lanyards, and while changing location from one work platform to another, he missed his step and fell.

The company retained me five days after the accident to produce a report of their Safety Management System as required by the Government.

Fig. 2.15 Double Lanyard

I visited the site three times, reviewed their scaffold design, erection and safety procedures, interviewed their supervisors and workers, and exhaustively analysed their plans and actions.

I then submitted a report with my findings and recommendations, which they had to implement before their stop-work order could be lifted.

My report was 23 pages long, with the following contents and page numbers:

A typical page of risk assessment violations noticed is presented as Fig. 2.16 as sample.

Also is shown the 3 by 3 risk matrix (Fig. 2.17) used to evaluate the risk for various activities from their likelihood and severity in Table of Fig. 2.16.

I also attached 17 pictures such as the two in Fig. 2.18, on each of which I pointed out various deficiencies.

3.1. Violations of Factories Act (Chapter 104)

FACTORIES (BUILDING OPERATIONS AND WORKS OF ENGINEERING
CONSTRUCTION) REGULATIONS

PART II : GENERAL PROVISIONS

Vio. No.	Cl. No.	Particulars of violation	Lkl.	Sev.	Risk	Control	Remarks
1	9.	Tripping and cutting hazards: (1) and (2) both violated in stairways and corridors. (Debris removal may not be contractual obligation of sub-contractor.) Sub-contractor's workers and supervisors habitually use the debris-choked stairways and corridors of the demolished building to reach the scaffold platforms and frames to carry out scaffold erection and dismantling, or inspection duties. This practice is extremely hazardous, especially where the stair handrails are missing or buried under debris	3	2	6	Workers not to use debris-cluttered stairs or corridors, unless special lifelines are previously provided along the staircase walls at every floor, before demolition work commences.	Photos 1 to 5
2	10.	Access to workplace: Ladders or other means of access to scaffolding not provided. Workers now climb on to the platform from the ends or outside frames. They also use the stairs, in spite of the demolition debris, as explained for Violation No. 1.	2	3	6	Provide proper access: Ladders attached to scaffold frames, movable as necessary from level to level.	Photos 6 and 7

Fig. 2.16 Sample extract from Table of Violations

Likelihood ↓	Severity		
	Minor (=1)	Moderate (=2)	Major (=3)
Frequent (=3)	3 (= Medium Risk)	6 (= High Risk)	9 (= High Risk)
Occasional (=2)	2 (= Low Risk)	4 (= Medium Risk)	6 (= High Risk)
Remote (=1)	1 (= Low Risk)	2 (= Low Risk)	3 (= Medium Risk)
Risk Index = Likelihood Score × Severity Score			
1-2 = Low Risk; 3-4 = Medium Risk; 6-9 = High Risk			

Fig. 2.17 3×3 risk matrix used to assess risk

Fig.2.18 Specimen photos from author's investigation of worker fall

This is one situation where you are paid to find fault! In fact, there is a strong third-party audit practice in Singapore, by which companies can find out their deficiencies and correct them before Government inspectors visit them and find them with much worse consequences. My task ended with submitting the report with my findings and recommendations, because the case had already been heard, and penalties meted out.

It must be mentioned that this accident occurred before the Workplace Safety and Health Act was introduced in 2006, and many workplace practices were tightened up for greater safety, gradually replacing the Factories Act.

2.12 FORENSICS IN VARIOUS ENGINEERING DISCIPLINES

2.12.1 Forensics in Structural Engineering

Columbia University describes structural forensic engineering in the following terms, [2.20]: *"Engineering investigation and determination of the causes of structural failures of buildings, bridges and other constructed facilities, as well as rendering opinions and giving testimony in judicial proceedings, often referred to as Forensic Structural Engineering, has become a field of professional practice of its own in the US.*

"With rapid economic development, increased design sophistication, more-and-more daring construction technology and accelerated project delivery came the proliferation of structural failures throughout the world. Several countries are reviewing and/or streamlining technical, business, and legal procedures modeled on US practices – with both their advantages and faults – which require expert consultants/witnesses in both the forensic investigation and in the ensuing dispute resolution."

Structures are the most common civil engineering products which fail in accidents.

Mechanics and structural engineering, and more recently biomechanics, are the author's forte, and he tries to accept assignments primarily in that area, computer graphics running a close second.

He has found that tools of the trade here should include:

- Strength and properties of materials, and structural testing labs;
- Mechanics and structures analysis tools, both classical and digital;
- Fair degree of familiarity with biomechanics analysis;
- Strong fundamentals of various factors that make structures and components stand up or fall down – not ignoring very basic statics and dynamics; and,
- Associated skills like computer spreadsheets and graphics.

Carper [2.21] lists the common causes of structural failures, as follows:

- Site selection and site development errors;
- Planning deficiencies – high client expectations;
- Design and/or construction errors;
- Material deficiencies; and,
- Operational errors, such as alterations, use change, inadequate maintenance, etc.
- Non-destructive testing (NDT) at the site and destructive testing in the lab are the common investigative methods in structures.
- In particular, concrete is quite amenable to NDT [2.22], as shown in Table 2.3.

2.12.2 Forensics in Transportation Engineering

On this, Hochstein [2.23] states, *"Forensic engineering in transportation matters requires the application of basic Traffic and Transportation Engineering principles in investigating, analyzing, reporting and testifying with regard to an incident on, or the design, operation, and maintenance of, a transportation facility, a building, commercial, residential and other public sites, pedestrian areas, etc. It may require services such as accident reconstruction, right-of-way appropriation, pedestrian fall down, driver behaviour, design defects, signing, seat belts, code reinforcement, vehicle mechanics, maintenance practices, etc."*

As traffic injuries and fatalities are the most common around the world, forensic analysis of traffic accidents is a very essential professional service, and many companies and experts thrive from it. Re-creation and simulation constitute a very large part of the investigation.

Table 2.3 Non-destructive testing methods

SR NO.	Information required (to determine)	Method / instrument available
01	Strength	A) Rebound hammer. B) Ultrasonic pulse velocity meter. C) Penetration probe. D) Pull-out method.
02	Concrete quality	A) Ultrasonic pulse velocity meter B) Penetration probe C) Gamma radiography
03	Concrete density	A) Ultrasonic pulse velocity meter B) Gamma radiography
04	Bar size and location	A) Cover meter B) Gamma radiography
05	Cover to reinforcement	A) Cover meter
06	Bar corrosion	A) Electro potential meter. B) Concrete resistivity

2.12.3 Forensics in Geotechnical Engineering

Saxena [2.24] opines: *"Forensics in the geo-domain encompasses an extensive array of topics with general emphasis in civil engineering and specific emphasis in geotechnical and related fields having geological, geophysical, geo-environmental, and structural applications. Mostly, it applies to failures after they occur when their application has prevented and/or identified failures prior to their occurrence. Furthermore, cases of analyses and evaluation of selected remedial measures, along with their effectiveness and economy, are normally subjected to judicial scrutiny."*

As geo-technical engineering deals with properties of soil and other natural materials of which we can find only limited data from which we must extrapolate to the rest of the domain.

Further, as these natural materials are susceptible to wide variations with complicated interactions with structures built in and on them, any accident which involves natural earth can get highly complex in its investigation.

Experts here should be really experienced in their work and be willing to face a lot of argument in their testimony. Singapore, induced by the Nicoll Highway collapse, has in recent years highlighted the importance of sound geo-technical analysis and design, and emphasised its role in accident prevention.

2.13 CONCLUSION

What has been presented here is the tip of the iceberg. Forensic engineering is essential to bring justice to the losers in an accident; but it is vital to prevent or reduce future accidents.

It needs special knowledge and skill, but more important, it needs a special attitude to convince a court with that talent. It is not a game to win or lose; it is a professional service to share for the good of the community – and it can be very fulfilling, beyond financial rewards.

2.14 REFERENCES

2.1.   _____, *Forensic Structural Engineering*, Retrieved, May 2013 from:
http://civil.columbia.edu/forensic-structural-engineering

2.1.   Carper, Kenneth L., "What is Forensic Engineering", *Forensic Engineering*, (Edited by K.L. Carper), 2nd Edition, CRC Press, 2001.

2.3.   Kielmas, Maria, "Forensic Civil Engineering Job Requirements", *Chron - Demand Media*. Retrieved May 2013:
http://work.chron.com/forensic-civil-engineering-job-requirements-11911.html

2.4.   _____, *The Kansas City Hyatt Regency Walkways Collapse.* Retrieved July 2013:
http://ethics.tamu.edu/Portals/3/Case%20Studies/HyattRegency.pdf

2.5.   _____, *About Disaster and Failure Studies*, NIST, USA, Retrieved July 2013:
http://www.nist.gov/el/disasterstudies/about.cfm

2.6. _____, *Rana Plaza Building, Bangladesh, 2013*, Wikipedia. Retrieved July 2013:
http://en.wikipedia.org/wiki/2013_Savar_building_collapse

2.7. Krishnamurthy, N., "Worker fall from mobile scaffold", *Int. J. Forensic Engineering*, Vol. 1, No. 1, 2012, p. 21-46.

2.8. _____, "Failure of a rolling scaffold", *Forensic Engineering in Construction – On-Line Edition*, Zallen Engineering, No. 13, march 2008.

2.9. _____, *Timeline of notable examples of progressive collapse*, Wikipedia. Retrieved July 2013 from:
http://en.wikipedia.org/wiki/Progressive_collapse

2.10. _____, *Hotel New World Disaster*, Wikipedia. Retrieved July 2013:
http://en.wikipedia.org/wiki/Hotel_New_World_disaster

2.11. _____, *Report of the Committee of Inquiry into the Incident at the MRT Circle Line Worksite that led to Collapse of Nicoll Highway on 20 April 2004*, Ministry of Manpower, Singapore, Vol. 1, Part 1, May 2005, p. 6.

2.12. _____, *Building Collapse Cases/Skyline Plaza at Bailey's Crossroad*, MatDL: Failure Cases Wiki. Retrieved July 2013:
http://matdl.org/failurecases/Building_Collapse_Cases/Skyline_Plaza_at_Bailey%27s_Crossroad

2.13. _____, The Chicago City Post Office, *Building Collapse Cases/Chicago Post Office - MatDL: Failure Cases Wiki*. Retrieved July 2013:
http://matdl.org/failurecases/Building_Collapse_Cases/Chicago_Post_Office

2.14. Johsnson, R.G., *Hartford Civic Center*, 2009,. Retrieved July 2013:
https://failures.wikispaces.com/Hartford+Civic+Center+(Johnson)

2.15. _____, *The Kansas City Hyatt Regency Walkways Collapse.* Retrieved July 2013:
http://ethics.tamu.edu/Portals/3/Case%20Studies/HyattRegency.pdf

2.16. _____, *Space Shuttle Challenger,* Wikipedia. Retrieved July 2013:
http://en.wikipedia.org/wiki/Space_Shuttle_Challenger

2.17. _____, *Richard Feynman*. Retrieved July 2013:
http://en.wikipedia.org/wiki/Richard_Feynman

2.18. Feynman, Richard, *What Do You Care What Other People Think? Further Adventures of a Curious Character, (as told to Ralph Leighton)*.

2.19. Krishnamurthy, N., Computer Applications in Forensic Engineering, *Proceedings of the Conference & Exhibition on Forensic Civil Engineering*, 23-24 August 2013, Bangalore, India.

2.20. _____, Retrieved, May 2013 from:
http://civil.columbia.edu/forensic-structural-engineering

2.21. Carper, Kenneth L., "Learning from Failures", *Forensic Engineering*, (Edited by K.L. Carper), 2nd Edition, CRC Press, 2001.

2.22. Madke, Rohit, *Forensic Engineering of Concrete Structures*. Retrieved July 2013:
http://www.scribd.com/doc/49515598/forensic-engineering

2.23. Hochstein, Samuel, "Ethical Considerations in Forensic Engineering", p. 235-237, Retrieved May 2013:
http://www.ite.org/Membersonly/annualmeeting/1985/A HA85C35.pdf

2.24. Saxena, Dhirendra S., "Forensic Geotechnical Engineering Application to Coastal Structures in Florida", *International Symposium on Geotechnical Engineering, Ground Improvement and Geosynthetics for Human Security and Environmental Preservation*, Bangkok, Thailand, 6-7 Dec. 2007, p. 183-194.

———

† This is a reprint of the author's paper presented at the Forensic Civil Engineering Conference and Exhibition held at Bangalore, India, on 23-24 Aug. 2013, by the Association of Consulting Civil Engineers (India), reprinted with permission.

3

USE OF COMPUTERS IN FORENSIC ENGINEERING[†]

3.1 INTRODUCTION

Computers have been with engineers for the last four decades so much that we now depend on them more than we depend on ourselves! Naturally, when it comes to investigating accidents and finding their primary and secondary causes, computers have a very important role to play. The 'Sherlock Holmes' lens of today is the computer through which complex mysteries can be resolved, details which were hidden may be brought to light, and so on.

Computers are extensively used for data management, statistical analysis, structural, geo-technical and other evaluations, and for computer slide or video presentations in court.

Needless to say, a forensic expert who wishes to use the computer in his work better be quite thorough with the concepts, their principles, and the details of computer applications, as well as on the spot manipulation of calculations and images. A forensic engineering expert looks weak if he has to call upon a computer expert in court!

[NOTE: Use of male pronoun or other reference would automatically cover the female equivalent except when either one is implied by context. – NK]

Lawyers on each side will have computer experts to debate with the testifying expert from the other side.

Judges also take the trouble to study up for the case, and may ask very probing questions which may stump an expert more than the complex analysis itself.

"How does the pixel resolution affect the accuracy of your results?" was one of the questions from the judge which I had to answer without making it sound like (or take as long as) a classroom lecture.

In my fifty-five years of active service, more than fifty of them have been spent with computers, starting with second generation mainframes right up to today's powerful desktop PCs, and from BASIC and FORTRAN to 'C' language (after which I stopped learning new languages), and sophisticated analysis packages.

During all that time, I had the opportunity to use computers to:

(a) Develop software for my own research and as consultant to private companies and the Government in USA, Singapore, and India;

(b) Publish numerous papers and lecture widely on computer applications, write a book on computer graphics, and contribute a section on finite elements to a book on implant dentistry; and,

(c) Consult for accident investigators (the term 'forensic' came later) and expert witnesses in USA with computer analysis of structural failures and accidents.

Now the computer is my strongest ally in my forensic engineering work.

I would not have been able to get thus far in – or contribute this much to – forensic engineering without it.

3.2 DATA MANAGEMENT

In our cyber world today, the computer is the storehouse and repository of all data on people, materials, equipment, processes, correspondence, financial matters, and everything else that is remotely connected with all our thoughts and deeds.

When an accident happens, the first act of the investigator should be to collect and secure any and all data available on the event, people and artefacts connected with it at the site. Time is of the essence. Computerisation of data is the most efficient means to securing data from loss, corruption, or manipulation. Computers are today the most effective way of documenting chain of custody/evidence to authenticate the expert's testimony.

3.2.1 Source of data

Data capture is fortunately easy and fast these days. The digital camera has revolutionised the visual medium. Almost every office and shop, most homes and condos of the rich – and in places like Singapore even government-built flats – many localities with unsavoury reputations, and other sensitive areas are watched by CCTV cameras. We get one chance at data capture, and we must grab all we can. We never know what will appear in which picture that may become the turning point in the investigation.

Normally the best source of data is the accident site itself. However, if you are called in to investigate long after the accident is over and the site has been cleaned up, a site visit may be fruitful only to orient yourself to the scale and ambience, but will not be practically useful. Of course, you must have access to other documented data from which you can proceed.

A few years ago, I was invited to investigate a formwork failure by the prosecuting Ministry of Manpower. It was already five weeks past the accident; the accident scene had been cleaned up and fresh work had started, and I did not visit the site. When asked in court, I explained that I had relied on the photographs, videos, and other records supplied to me by the lawyers for my investigation.

The beauty about computerised data management is its reproducibility and accuracy, almost unaffected by who does it, so that there could be no question of bias, and once thoroughly checked during the first round, no question of human error. It makes for a good evidence trail.

In my own forensic preparation, I utilise the computer for all data management from storage, search, or data mining from the web.

3.2.2   Data and Testimony

Computerisation of data is also useful for further analysis, for transfer to other parties in the loop, and for presentation during testimony, or for validity check of the other side's testimony.

In one exchange with a witness involving the possible failure of a temporary structure due to overloading,

I had a slide of the construction site projected on the screen and pointed out that contrary to defence claims, there was no "designated area" marked out for the heavy load to be deposited by crane, as required by design regulations. It had also not been marked on drawings.

In another court appearance, I had occasion to provide technical explanations on the various stages of how the victim fell, through the CCTV video of the accident moved frame by frame.

3.2.3   Nature of computer evidence

Computers are as necessary to modern forensic work as they are suspect in their integrity.

As most people know, with a computer, any desired text or numerical output can be developed, any existing image can be altered, and any new desired image can be created. Because computer processing is invisible to the human eye, courts are very strict in accepting evidence based on computer-created text and images. The only two ways out of this problem are:

(1) Pre-certification by the source and chain of custody of data, as with Government documents; and,

(2) Post-authentication by specialised and accredited computer labs, based on meta-data and very, very tedious and expensive tests.

The same scepticism may also apply to impressive results based on computer analysis. As it is practically impossible to directly check the accuracy of a computer output in say a finite element analysis, the only way to is to independently run the analysis on the same or equivalent software with the original data and compare the results.

If the two results differ drastically, it gives the opposition an opening to pick a hole in the input; it also gives them a chance to argue that the submitted results are 'mis-represented' (meaning manipulated) and to demand an explanation for the difference.

3.3 COURT PRESENTATIONS

Today's courts are a far cry from courts of a few decades ago, although decades-old practices continue in some nations without modern infrastructure. Almost everything in courts is now computerised from record keeping and scheduling, to facilitating visual graphic presentations, both still and video.

Visuals are unavoidable in forensic engineering presentations. Earlier, witnesses would display visuals through flip charts or overhead transparencies.

Huge photo enlargements were made of photographic evidence so that the judge, lawyers, witnesses, the media, and other key spectators can see and understand the evidence clearly. Then 35 mm slides projected on a screen served well for many decades. The current popular mode is LCD projection from computers, most commonly with slides like by MS-PowerPoint.

Small scale models of accident sites and artefacts (or conversely large-scale models of very small items) would be used to give all parties a clear idea of relative locations, sizes, and functioning of different components.

In traffic accidents, models of vehicles involved and the street and intersection layout would be essential. Mannequins would be used to point out body parts and postures if human beings were involved.

These strategies are still used in court to give a 3D view of things and actions. In a case I testified on the failure of a rebar grid, my lawyers had a model made of the cage to point out various parts and locations.

More and more, rather than physical models, photographs or computer renderings of physical models or of the objects themselves are accepted for testimony, with mutually agreed upon stipulations on their admissibility for presentation, and for further proof of authenticity and representation as occasion demanded. Computers do speed up presentations, and make the proceedings much more interesting and meaningful than otherwise.

3.4 MATHEMATICAL ANALYSIS

Current use of mathematics (in my area of speciality structural engineering) ranges from basic analysis by classical principles (even hand-worked with the aid of calculators) to highly refined matrix and finite element analysis, vibration and earthquake analysis, etc. by computers.

Validity of computer analysis is critical in forensic engineering because the investigator cannot simply do his favourite thing with the computer and expect it to be accepted in court as gospel. He must anticipate arguments from the other side, try various computer models and be able to justify his particular choices, and recommend one finding over the other with credibility.

3.4.1 Pitfalls in computer packages

Along with the advantages of computers come certain limitations.

Any computer package is a product of one or more minds, its power and validity depending only on the experience and expertise of those minds.

Most computer packages today are so massive and complex that a user cannot look into its programming guts, and if he could he would not understand much. At the same time, these are the very reasons the user should clearly understand the following:

- Particular principles or theories which have been adopted in the software;

- Assumptions on which the programme has been written;

- Criteria and limitations to various modelling options in the input stream;

- Transformations and approximations used to get and present the results; and,

- Changes from one version of the package to the other.

Packages offer choices of 'models' and idealisation options for the user to choose from, and that is where the novice or over-confident user can go wrong.

The craze for the latest version of a popular package overlooks the truth that the user might not have exploited a fraction of the capabilities of the previous version! Where human judgement is necessary, automation may not always give the right option under the circumstances.

The final big hurdle in computer software use is that the numerical output is so voluminous that nobody can wade through them and extract the most critical values he needs to decide between alternative designs or to proceed for further analysis. So we have to depend on the summaries and graphical representations as decided by the programmers.

In other words, unless one is careful, everything except the input of the bare essentials is out of the hands of the user of the package!

3.4.2   Matrix and Finite Element Analysis (MFEA)

MFEA probably leads computer use in forensic civil engineering investigations.

These days, most finite element analyses are massive jobs, running to millions of equations, with powerful software and hardware to match.

Because of their ability to probe deep into the behaviour of solids, the expense involved may be worthwhile in major accidents.

The main characteristics of MFEA are:

- Today's desktop computers can handle huge problems within a short time.

- Interactive graphics will not only check the validity of your input but can actually facilitate it through helpful graphics and text entry wizards.

- Data validation and error correction guidance are generally built into most software.

- Results are displayed graphically so that you can confirm the reasonableness of the answer immediately.

- With a few keystrokes you can change essential parameters of the computer model, and see their effect on the results right away.

Subject to the caveats mentioned earlier, computer results are the best legal ('forensic') evidence, to the extent that anything the expert has done should be reproducible and can (and often will) be questioned and checked by anybody in the loop, particularly the other side.

3.5 COMPUTER ANALYSIS IN FORENSIC ENGINEERING

Apart from the modelling of the structure itself, questions will arise on the exact site conditions of material properties, field connections, supports and loadings. That is where deviations from design and intended use occur.

The analyst would have to try various combinations of parameters before the result gives indications towards possible causes.

A couple of recent examples from public domain, and one case study from personal experience will be reviewed.

3.5.1 Collapse of steel bridge on I-35W at Minneapolis, USA, 2007

About 6:05 p.m. on Wednesday, August 1, 2007, the eight-lane, I-35W highway bridge over the Mississippi River in Minneapolis, Minnesota, USA, failed catastrophically. The central 1,064 ft. (324m) long deck truss portion of the bridge collapsed, with adjacent sections of approach spans. Thirteen people died, and 145 people were injured. (Figure 3.1 displays the collapsed bridge with inset of a view before collapse; Fig. 3.2 shows further details.)

Fig. 3.1 I-35W bridge, before and after

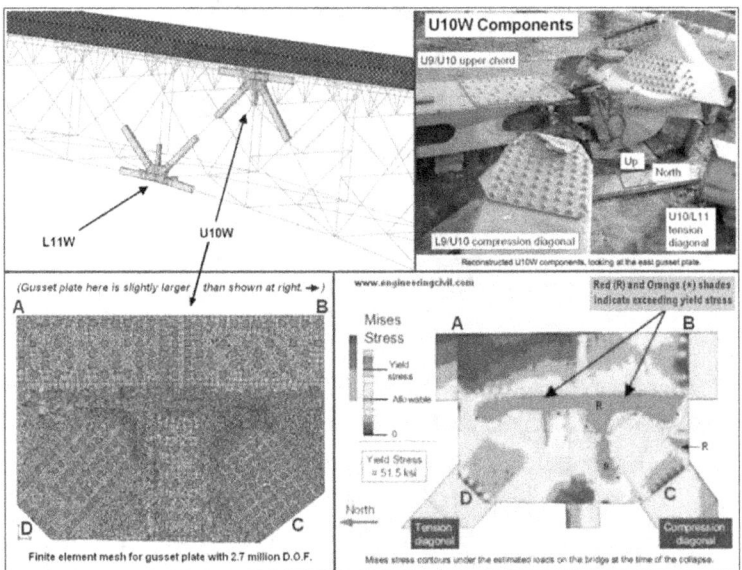

Fig. 3.2 FEA of bridge collapse on I35-W at Minneapolis, USA

Gusset plates were suspected to be the failed element; Figure 3.2

top right depicts one of them.

To confirm and quantify their role in the collapse, consultants Simulia using the ABAQUS package conducted an elaborate finite element analysis (FEA), [3.1] of the entire bridge and of two of the failed gusset plates marked in Fig. 3.2, top. The other three parts of the figure show the actual failed gusset plate for the upper joint recovered from the river, the 3D finite element mesh with 2.7 million degrees of freedom (– so fine we cannot see the mesh!) and the von Mises stresses clearly marking the failed zones.

3.5.2 Fire effects on beam and slab floor, 2001

Apart from stresses and deformations, FEA can also determine effects of temperature, vibration, impact, creep etc. Figure 3.3 shows the forensic analysis of fire effects on World Trade Center – 5, part of the 9/11 attack, [3.2]. This building was not directly hit by terrorists, but raging flames in the Twin Towers spread to WTC-5 and caused localised collapse of the top five floors.

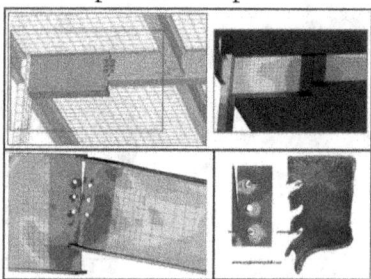

Fig. 3.3 Fire effect FEA

The FEA solution by ABAQUS package was a heat-transfer thermal-stress analysis and its results agreed quite well with the observed fire damage. The structural model included spray-applied fire insulation and concrete slab on steel framing, as in Fig. 3.3, top left, for a shear connection. The top right figure shows the temperature distribution near the connection after two hours of the fire exposure.

Bottom left part of Fig. 3.3 shows von Mises stress distribution and deformations after two hours, and bottom right part shows comparison of FEA and actual forensic evidence of the torn plate, confirming bearing failure at the bolts.

The analysis confirmed that the insulation delayed heat transmission to the steel, and the concrete slab served as a heat sink as anticipated.

3.5.3   Collapse of rebar support system

A major finite element application to forensic civil engineering came my way when I was invited to investigate the collapse of a rebar support system for 3 m and 5 m reinforced concrete slabs. I have touched on some aspects of it in another paper, [3.3].

I did scores of MFEA, first to check the analysis and design submitted by the designers and contractors, and next to try various possible support and loading scenarios to explore the failure.

Figure 3.4 displays some of the support system configurations I analysed, using commercial software – thanks to vast computer facilities offered by a friend. Although the biaxial symmetry was a compromise for detailed analysis to save unduly long runs, a few coarse mesh runs were made on the full domain for unsymmetric loadings to get an idea of how big the differences would be.

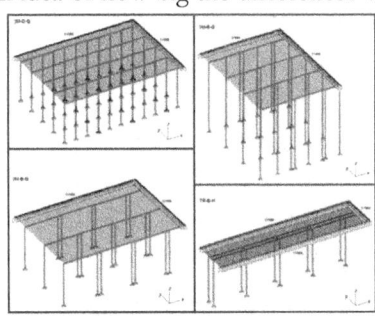

Fig. 3.4 Configurations analysed by FEM

Figure 3.5 depicts one of the larger computer models of the subject failure. This was just one quadrant, so that the actual structure would have had four times as many joints and members.

Finite element analysis of any practical scope is not a one-person job. Of course the lead investigator has to conceptualise and lay out the model and its 2D or 3D configurations, material properties, support conditions, loadings etc. But the preparation of input is quite time consuming and highly software dependent, and better done by a person trained in that particular package.

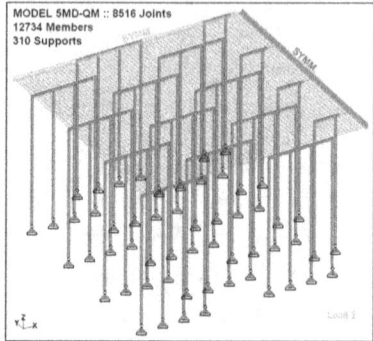

Fig. 3.5 Computer Model of Quadrant

The investigator should know enough about the package to ensure that the data developer is on the right track, ask intelligent questions and offer sensible suggestions. Accordingly, I used the services of an experienced assistant to prepare the input, carry out the runs, and document the output, all under my supervision and instructions.

A key answer I sought was the factor of safety against failure under worst loading conditions. My analysis showed that the minimum factor of safety (F.S.) for the 5 m support, which was more critical than the 3 m support, was 1.2 under design loading, but less than 1.0 for the extra loading. Further clarifications could not be made purely on scientific grounds or factual basis.

All I could do then was to present credible scenarios of failure. As an expert witness, I had done my job, given my testimony, and was soon out of the loop.

Eventually, the company accepted liability for deficiencies in design and construction and paid a heavy fine, but ultimately no individual was charged specifically for the failure under the regulations applicable at the time.

There were of course, some arguments on what constituted

safety in temporary structures. Defendants felt that F.S. of 1.5 as for permanent structures would be enough, but I held to my opinion that we needed a minimum of 2.0, which had also been recommended by the Committee of Inquiry for another high-profile case earlier. Again, at the time of collapse, the low load factors were admissible, so my recommendation was just that and not grounds for a charge.

In due course, when I was assigned the responsibility to develop the design section of the national Formwork Code Committee, I led the move to set minimum safety factor for formwork at 2.0. This requirement is now built into the Code for any method of design or testing.

This is an example of how forensic engineering highlights weaknesses in design or construction procedures, and facilitates their rectification to help future users, regardless of whether earlier mistakes were penalised or not – such laws cannot be retroactive!

3.5.4   'Playing around' with computers

Forensics is actually being a detective, going backwards in time, trying to find out the cause of some mishap that has already happened. In order to succeed you need a tool that can help you try out various scenarios and check if, how, and how well any one scenario fits known facts best.

For this guessing exercise, the computer is the ideal tool. Before computers it used to be a mind-game, a paper task, or a lab experiment, all complex and/or time consuming. The computer eliminates both deficiencies. It is precise, fast, and relatively inexpensive.

Because of the ease and speed of computer analysis, whenever I investigate a new problem which requires computer analysis, I use the model I have created to not only answer the questions asked but also to 'play around' with it trying various other scenarios, as much to understand my solution better, as to learn anything more that can be gleaned from it.

This approach has given me some valuable insights. If I had not found out some unexpected problems during my playing around, an accident or failure might have happened sooner rather than later – should I then call such playing around, 'preventive or pro-active forensics'?

Laying steel sewer next to R.C. tunnel:

One such episode occurred when, while at Auburn University (1967-1975) I was consulted on the safety of a (then) 60-year old reinforced concrete sewer pipe during the laying of a new adjacent steel culvert, in a neighbouring city. (Figure 3.6, a, b, c.)

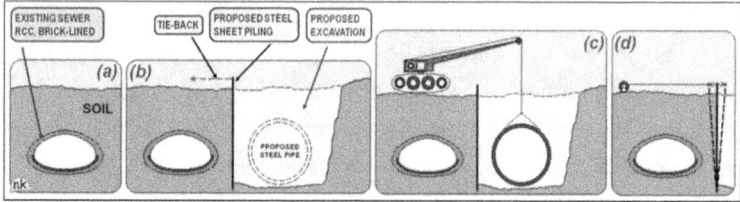

Fig. 3.6 Playing around with computers on R.C. sewer

I chose FEA for it – a daring move at the time – and showed that when the tractor crane hauling the steel pipe rode over the buried concrete pipe the old sewer might just make it through. I recommended placing spreader steel plates under the track and driving slowly.

Then, over the phone, the client asked, what if the sheet pile tie-backs slacked off during use?

No problem; I input a top displacement slack of half an inch (the specified tolerance) into the data and came out with the result that it would decrease the maximum stress by about 5%.

That ended my assignment. They were going to launch the project the next morning.

But overnight, the thought struck me that the tieback might as well be over-tightened during the process of erection and adjustment, rather than slackened.

So I sent word to the client and asked them to hold off the culvert laying by a day. Next morning, I rushed to the computer centre and waited for it to open.

The re-analysis for half-inch over-tightening gave me 10% overstress, which would almost surely have destroyed the old sewer. I called the city fathers and advised them to make sure that there was no over-tightening.

Also, I had second thoughts on what the tractor load would do to the soil under these fluctuating tie-back loads. So I recommended that they avoid loading the old sewer at all, and reach the excavation from the other side. It meant some additional time and expense, but they appreciated the forewarning – a failure would have meant court cases and (in those days) hundreds of thousands of dollars worth of claim settlements.

3.6 FAILURES DUE TO SOFTWARE

3.6.1 Sources of computer-related failure

We will not be discussing the misuse of the computer itself, such as spam, virus, hacking, identity theft, or other aspects of cyber crime, which is a separate field in itself. Computer hardware error also is very, very rare; the Pentium FDIV bug of 1994, [3.4], giving error in the fifth significant digit during division is the only one in recent memory.

But there have been many errors or limitations in computer software which have gone unnoticed for long and found to be the reason for many wrong analyses and bad designs . Recent history is full of accidents and failures due to misuse of computer programs.

The latest software error reported by Confidential Reporting on Structural Safety (CROSS), [3.5] reads thus: "... *a current package for pad foundation design has no factor of safety against overturning, returning a 'pass' for unfactored loads and a utilisation ratio of 1. Furthermore it has been noted that the same program returns a 'pass' without checking the bending capacity of the base in hogging. ... The suppliers ... have undertaken to correct them at some point but there may be many pad foundations in use which have been incorrectly designed using this program.*"

The Editorial on it comments thus: *"... there is concern about reliance on computer output and there should always be a check to ensure that results are sensible. In simple cases such as a pad foundation this can be an approximate manual calculation."*

Software errors are also infrequent these days as reputable software houses have very stringent check and test procedures. In any case, to avoid liability problems, most software contracts carry a disclaimer clause which in effect says that the supplier will not be responsible for any damage or loss caused by the use of the software – *'Caveat emptor'*, "Let buyer beware!"

Currently therefore, most accidents involving computers in accidents can be traced to:

(a) Incorrect application of correct computer programmes;

(b) Wrong data input; or,

(c) Wrong output interpretation,

– all of which could be grouped under 'software misuse'.

In particular, MFEA was and continues to be an art and a science. While the number crunching and more recently animated colour graphics are the computer's main contributions to this powerful tool, two areas are – or at least should be – still the domain of the human mind:

(i) Modelling of the problem; and,

(ii) Interpretation of the results.

Most MFEA-related failures happen because of deficiencies in these two areas.

3.6.2   The computer trap of over-reliance

The very speed and power of the computer are also its built-in danger. Computer users soon get sucked into its sway, and blinded by its glamour. The hardware and software gradually take over the user's thinking, lulling him into a false sense of security and trust – until disaster hits.

Computer results can never substitute for understanding structural behaviour. The engineer should know the approximate 'ball-park' answer before going to the computer and must be able to distinguish an accurate solution from one that is absurd but appears precise. [3.6]

The algorithms of automatic input preparation and output display are the product of someone else's brain. By now, almost everyone who uses MFEA has lost touch with what goes into input and what comes out as output. Many are using very powerful finite element packages for very complex structures and phenomena, without the least idea of the element being used, the latest theory embedded, the specific criteria applied, or the possible effect of modelling on the results.

'Wizards' – those smart routines which ask for just a few numbers and clicked choices, develop an entire mesh, run the analysis, and wrap it up with nice colour graphic results – are like some magic or voodoo that gets what you want fast and effortlessly, no matter that you don't understand what, how, or why. Serious professionals would at least spot check their creations.

Until I gained enough know-how, I always tried to run every FEA problem with at least two and preferably three different meshes, not to eliminate the 'discretisation error' (because you really cannot eliminate it altogether) but to know the magnitude of potential error in my result so that I can extrapolate to the true value for the continuum, as in Fig. 3.7.

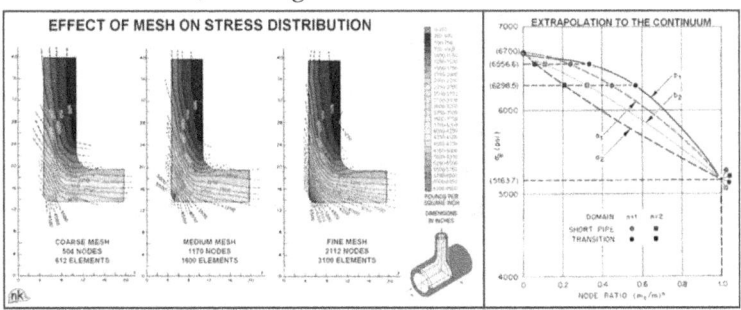

Fig. 3.7 Mesh effects and extrapolation to continuum

In the late 1960s, this was standard practice for serious work. Without offering FEA error bounds and an estimate of the corrected value, I doubt if the U.S. Atomic Energy Commission would have accepted my recommendations, [3.7]! Even now, I do not trust my own single mesh answers, unless I have similar mesh experience or available literature has equivalent results.

Today, FEA tools have become smarter. The discretisation problem is mostly pushed out by sheer brute force of too many degrees of freedom. Even so, without the awareness of modelling effects, a FEA package could be like a knife in the hands of a child.

A finding in a paper [3.8] by Bella and Liepins is worth quoting (– better than just my say so!): *"Today's engineering graduates are well-versed in the matrix structural analysis methods that form the basis for computer analysis, but they are weak in the classical hand methods that allow approximate checks of finite element methods and develop a feel for structural behaviour."*

To be fair however, we should not condemn all computer use outright. Surely, many users have studied the theory of MFEA in college and maybe even done projects with it. But once they come out of college, life takes on a different hue, getting results out 'yesterday' is important, and so youthful curiosity and healthy scepticism take a back seat.

Further, stupid mistakes can be (and have been) made even with manual calculations and graphics. The only differences with computers are that:

- The tool was developed by someone other than yourself in a process which you did not share;
- Once inside the computer, too much happens too fast;
- The entire process is invisible; and,
- You cannot check back the output and locate what went wrong and where – you can only check your input.

That is why computer applications need extra care, as some examples will illustrate.

3.6.3   Hartford Civic Centre Arena Collapse, 1978

Main details of this case have been covered in another paper of mine, [3.9]. Here I will touch only on the problems created by improper use of software by the designers.

The arena roof was a three-dimensional truss designed for the first time by a three-dimensional matrix truss analysis computer programme.

Erected in 1973, it served without incident for five years. But on the night of 18 January 1978, it collapsed under a snow storm. Just by good fortune, there was no one in the arena at time of collapse, [3.10]. (Figure 3.8.)

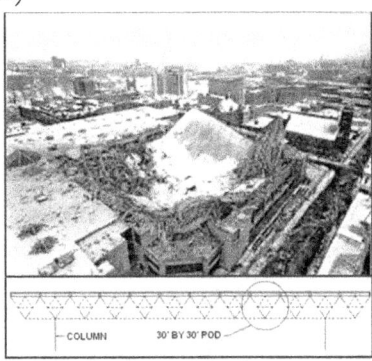

Fig. 3.8 Hartford Civic Center collapse

While most of the blame went to the fabricators who shifted the welded connections at the roof by a few centimetres, it was also discovered that the designers too did not realise the limitations of their computer programme, and did not heed the implications of the large deflections the roof experienced during the erection stage, blindly trusting the computer design for the strength and stability of the roof structure after erection.

The small shift of the designed connection had reduced its strength to less than a tenth in one case, and less than a third in another. Investigation after failure brought out the following limitations of the new computer programme which led to the large drop:

- The buckling mode of failure, to which roof design was extremely susceptible, was not considered in that particular computer analysis and thus not provided for.

- Incorporated into the computer model were some fundamental assumptions about end conditions on certain frame members, which turned out to be grossly oversimplified.

- Connection details were difficult to incorporate in the computer model.

As a combination of these factors, and the users' over-reliance on computer analysis with an imperfect model, the seriousness of the fabrication changes to the connections was not apparent.

The reactions to the collapse were swift and very far-reaching:

- It shook public confidence in space truss roofs, and even more so, in the new-fangled computer analysis of structures.

- President Gerald Ford ordered water load testing for a similar roof in Michigan.

- Engineers and architects tempered their reliance on computer models to cut down the structure to bare minimum, leaving no redundancy or margin for error.

- It focussed on the need to examine what exactly a computer program did, and how a structure needed to be modelled for computer analysis to reflect the desired design, as well as to be re-analysed when the construction/erection conditions change.

- It highlighted the need to watch out for warning signs such as deflections or deformations during the construction stage much higher than predicted in the design.

3.6.4   Sleipner Off-Shore Platform Collapse, 1991

Decades after Hartford, computer confusion still remained with us! The Sleipner platform to produce oil and gas in the North Sea, standing over 82 m of water on a concrete gravity base structure, sprang a leak in one of the supporting cells on 24 August 1991 during erection, crashing the whole platform and causing an economic loss of $700 million, [3.11]. (Figure 3.9.)

Why? After all they used among the most popular finite element package in the world at the time, NASTRAN! Yes, but the problem with finite elements are not how good a package you have, but how good you are <u>with</u> any package you have!

Fig. 3.9 Sleipner Off-shore Platform, before collapse, and schematic

Post-accident investigation traced the error to inaccurate finite element approximation of the linear elastic model of the supporting cells – simply put, the mesh was too coarse. The shear stresses were underestimated by 47%, leading to insufficient design. More careful FEA made after the accident, predicted that failure would occur with this design at a depth of 62 m, which matches well with the actual occurrence at 65 m.

How fine a finite element mesh must be cannot be decided by the 'wizards' that accompany a modern package to supposedly make your job easier, but actually taking away your initiative.

3.6.5   How bad can the variations be?

The 113 page paper [3.12] by Professor Emkin, Founder and Co-Director of Computer Aided Structural Engineering ("CASE") Center at GeorgiaTech is very revealing. Figure 3.10 shows the 67 story reinforced concrete building analysed for forces, moments and deflections, by the following five models, and plotted results along certain lines:

1.   FEA – Full Finite Element Model, with FE Floor Slabs

2.   RBPCD – Rigid Body Plane floor membrane, including Column axial Deformations

3.   RBPNCD – Rigid Body Plane floor membrane with No Column axial Deformations

4. RBSCD – Rigid Body Solid floor including Column axial Deformations

5. RBSNCD – Rigid Body Solid Floor with No Column axial Deformations

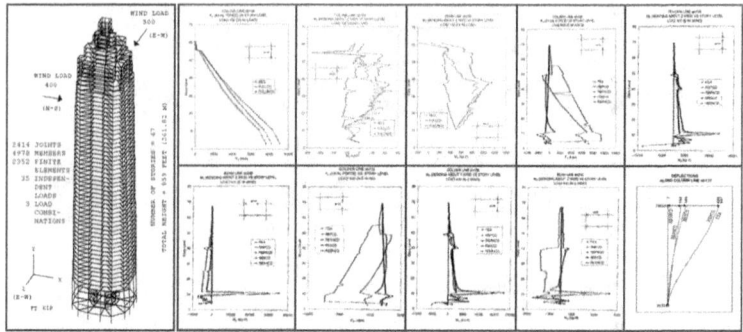

Fig. 3.10 Emkin's RC Building analysis of computer solution accuracy

The exact values in the tiny pictures may not be readable. But the wild and wide horizontal swings in the plotted quantities in the ten charts from the five methods should be impressive enough about how the choice of computer model can govern results from computer MFEA.

I have highlighted (more basic) variations in computer modelling in another paper, [3.13].

3.6.6   Case Study from Geo-technical Engineering

The computer is impartial in its distribution of its woes to engineering topics other than finite elements and structural analysis. Geo-technical engineering is particularly susceptible because soils have so many variable properties that the user needs expertise and extra care with the input parameters and the failure model chosen. The following example of 2004 Nicoll Highway Collapse in Singapore is an example. (Figure 3.11.)

This accident has been discussed in some detail in another paper of mine, [3.9]. The computer problem identified here is the wrong choice of geo-technical modelling option in the Plaxis software package used. The designers had chosen Method A, while the prevailing soil conditions required the use of Method B, which would have predicted much greater wall displacements and moments than Method A, as measured at site.

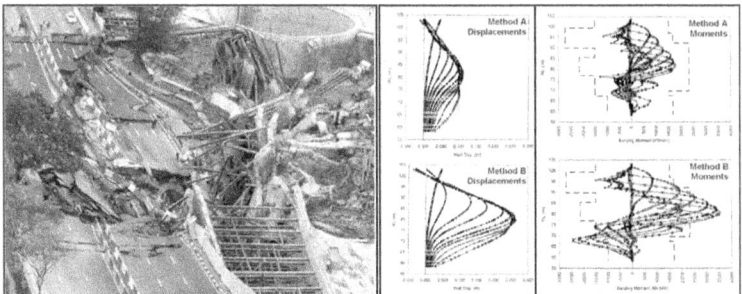

Fig. 3.11 Nicoll Highway wall collapsse, Left: The scene, Right: Plaxis results

Figure 3.11, right, shows the relevant charts from the report of Committee of Inquiry, [3.14]. Again, all we need to compare are the spread of the curves in the top (Method A) and the bottom (Method B) charts for displacements and bending moments to see the shocking truth that choice of the wrong model under-predicted the effects by about 50%, leading to the major tragedy

The fact that daily measured displacements were consistently and considerably more than those predicted by the design should have clued the personnel responsible to re-examine the model, but it did not.

3.7 SPREADSHEETS

3.7.1 Advantages of spreadsheets

Computer spreadsheets such as Microsoft Excel are a boon to forensic investigation, because:

- They can handle vast amounts of data – 1 million rows by 16,000 columns in Excel.

- They can exploit most of the advantages of computer data management listed earlier.

- They are always ready, fast and interactive.

- They will accept all kinds of data and can be easily programmed to arrange, sort and manipulate them, compute required answers, and display results in various tabular and graphical forms.

- The data validation aid available can help avoid entry of wrong data, and highlight inconsistent items in lists.

- They will let us change any data at any stage of analysis and immediately modify all affected quantities, as well as alter the related graphics to reflect the change made.

- They will do basic statistical analysis on data and results and display related graphics.

It suits the technique of experimenting and learning by playing around. I have had a lot of fun with it, and am using it more and more to satisfy my curiosity as well as to do my tasks.

Needless to say, spreadsheets can also be risky, partly because they are so easy to use. The biggest danger is that unless there is rigorous documentation and/or locks on the cells which embed formulas or conditional formatting, an accidental entry into one of these cells can wipe out the entire functionality.

Worse yet, it may let the programme continue to work, accepting wrong data into the loop due to the accidental entry, and eroding the integrity of the results, without obvious signs that something is wrong! The user must keep checking back with standard results occasionally to confirm that no accidental corruption of the programming has taken place.

3.7.2   Statistical Analysis

Spreadsheets like MS-Excel have a number of basic statistical functions such as Anova, Correlation, Descriptive statistics, Histogram, F-test and t-tests, and some limited regression analysis. Combined with graphical representations of data by bar-charts, pie-charts etc., I have found them quite adequate for my forensic engineering activities.

However, at a higher level, where the entire outcome of a case may depend on statistical findings and recommendations, some professionals may not be too happy with Excel's accuracy or scope, and then you would have to go to more sophisticated statistical packages like SPSS.

3.7.3   Parametric Studies

Frequently it would become necessary in forensic engineering for parametric studies to find the effect of variations of different quantities affecting some critical outcome. Although this can be done with many general purpose analytical software, more often than not one would have to port the results from them to a graphics software for the plotting.

Spreadsheet software like MS-Excel is ideal here, because not only can it quickly generate computed results for multiple sets of variables, but also can it plot a variety of graphics from any required set.

As and when the data changes, the graphics also changes simultaneously, giving a very powerful interactive decision-making tool for the forensic engineer.

In a recent investigation involving the fall of a worker from a mobile scaffold resulting in a head injury, the question was raised what the height of the guard-rail should be for a worker to safely lean on in normal course of work. (Figure 3.12, left.) This analysis has also been mentioned in another paper [3.9], and discussed in detail in a journal paper, [3.15].

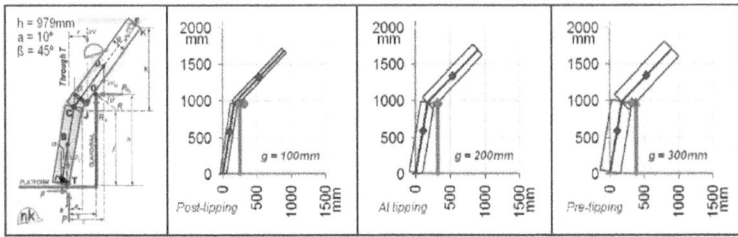

Fig. 3.12 Computer Analysis for Falling Over a Guard-rail

I conducted a parameter study of the characteristics of workers of various girths falling over guard-rails of different heights. (Figure 3.12, right three.)

The charts show for instance that with a guard-rail which happens to be 979 mm high, a person leaning towards the rail at 10° and over the rail at 45°, will fall over if he is 100 mm in girth, just balance on the rail if he is 200 mm girth, and will be safe from falling if he is 300 mm girth, because his centroid will be outside, on, and inside the guard-rail – so fatter the better!

The point is, once I had set up the model, how simple it was to incorporate the equations into the programme and come out with interactive graphics – quite thrilling to see the stick man move at my command and tell me if he will fall or not! It turned out to be a good learning experience about falling over a hand-rail.

3.7.4   Presentation of data and results

The old adage that *"one picture is worth a thousand words"* can be exploited to full effect with computer graphics software, starting with bundled software like MS-Paint to sophisticated stand-alone packages like Adobe-Photoshop.

You can do wonders with simple software, even with the built-in capabilities of spreadsheets being limited to pictorial representation of data or results and not useful in creation of new images.

Spreadsheets give the facility of producing bar-charts, pie-chart, and *x-y* plots simultaneously with the computations, in clear and colourful formats. (Figure 3.13.)

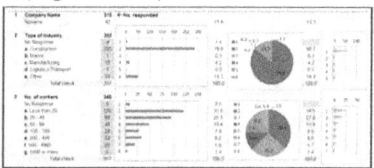

Fig. 3.13 Spreadsheet Graphics

These are a blessing when the forensic engineer wants to present the effects of some factor varying over a range, or the distribution of some critical value among various segments.

Statistics presented pictorially are not only prettier but also more powerful in its immediate impact on the audience. Comparisons are easier visually through charts than by numbers.

3.8 COMPUTER GRAPHICS FOR HIGHLIGHTING AND MEASURING

Although the computing capacity of computers is indeed what lifted mankind to a new level of professional sophistication, it is the GUI (Graphic User Interface) that really exploited the analytical capabilities of the machine.

Engineers from previous generations are lucky that they would have gone through a few manual drafting courses as students. Today, with plotters doing the drafting, sketching and drawing are lost arts.

In addition to my drawing courses, I even had some brief coaching in freehand sketching from an artist.

But equally important has been my fascination with graphics as a skill and hobby, ending up with my writing a book on computer graphics, [3.16].

I soon found that the ability to sketch gave me a powerful advantage over those who did not have the facility. In my teaching, writing, and consulting on technical topics, I freely use graphics, especially 3D representations such as isometric and perspective, to great effect.

That computer graphics can be a useful tool in accident investigation and forensic activities may need no explanation. But the variety and versatility of its uses can be amazing.

3.8.1   Forensic proof

Photographs can be used to prove your contention or disprove the other side's contention in a court. As already mentioned, digital pictures have to be authenticated before they can be accepted as evidence; interestingly, old-fashioned negative-based prints (with original negatives) are still good for evidence because the molecules in the negative cannot be altered!

In one of my cases, a contractor claimed he had bound his rebars with wire at regular intervals according to regulations or standard practice. To disprove his contention, I showed a photograph of the area (Fig. 3.14), which clearly depicted the sparseness and irregularity of 'w' marks denoting the bindings.

Fig. 3.14 Graphical enhancement of photographic content

The same photo also shows horizontal bars (B) links (L) and ties (S) with their centre lines marked, documenting their sag, out-of-straightness, lack of verticality etc. The 'w' and centre line marks draw the viewer's eyes to the essential elements. (You would have to submit the unmarked originals to the court.) While these by themselves could not prove that the lapses led to the collapse, at least they could cast doubt on the contractor's credibility.

I have used comparisons of two photographs of wire bindings on bottom rebars (shown by light coloured circles and stars in Figure 3.15) to argue that the near-perfect conditions of a lab test did not simulate the actual conditions at the accident site.

Fig. 3.15 Wire binding for rebars, Left: At site, and Right: In lab test

Regardless of the scale of the two photographs (which at the time I meant to represent nearly the same extent) the orderliness and closeness of bindings in the lab specimen are in stark contrast to their randomness and rarity in the site shot.

3.8.2   Forensic Computation

What do you do if you need the dimension of some item, but you have only a picture of it, and no other information?

You can use the geometry of the particular item in relation to one or more items in the picture whose dimensions are known or can be estimated within some tolerance. I have used this technique more than once, but it takes solid science and lot of effort to convince the court.

Although I have included this topic under computers, a forensic engineer can do it on an enlarged print of the photograph also, by actually measuring on the photograph in millimetres instead of computer pixels, and arrive at the same findings, as they used to do in the olden days.

It is just that it is much easier with computes, and quite simple and fast to prepare visuals for the analysis and result presentation for your analysis.

(a) Bent rod deflection:

In one investigation, I had to determine what cushioned an injured worker's fall so as to prevent him from serious injury which should otherwise have happened according to basic dynamics of falling, [3.15].

The helmet he wore (without a chin strap) had fallen first and been crushed by the falling scaffold. But the worker survived.

All I could find in the photograph was a bent rod forming part of the guard-rail. The worker's body must have hit the rod and bent it, absorbing much of the kinetic energy. (Figure 3.16.)

Fig. 3.16 Deflection of rod

The figure shows the top of the fallen scaffold with one helmet crushed under the far leg, the helmet of the second worker who escaped injury lying close to it. The bent rod A'B, and the desired deflection are marked on it. The near end of the rod, tied by wire to the top rail (which itself is illegal) has slid down from its original position A to a stop A' at the vertical mid-rod.)

To check out the impact force, I needed the deflection δ. But the rod was long gone from the scene, and God knew what happened to it.

So I went to work on the photograph on my computer. These days you get most pictures in their soft copy version and you don't have to scan them. But my lawyers had only hard prints, so I had to scan them into my computer first. Then I noted that δ spanned 39 pixels in the photo.

To find how much δ was in mm, I needed a reference length. Generally, if and when I take a photo where I do not know one or more of the dimensions, I lay a 6-inch scale (which a forensic engineer is supposed to carry around with him) or at least a ball-point pen or some personal article of mine for reference. But this was not my photo.

However, I knew the plan dimensions of the scaffold as 1.8 m by 1.2 m. The photo perspective decreased the apparent dimensions as the object receded farther from the camera, as shown for the 1.2 m side bar decreasing from 343 pixels to 198 pixels. The 1.2 m at mid-length of AB could therefore be estimated as being represented by (343+198)/2 or about 271 pixels.

Then, deflection at mid-length could be calculated as: $1.2 \times 39 / 271 = 0.17$ m or 170 mm. I was able to confirm that the energy required to bend the rod to this deflected shape had reduced the final impact to his body, in a way 'cushioning' his fall and preventing fatal injury.

How accurate is this deflection estimate? I have taken linear behaviour for the perspective, which is not quite true: mid-span will not be mid way between the ends in the photo – so I have not been too precise. But given time and incentive (like the need to prove more rigorously) we can get all these factors reflected in the calculation. However, for a first order estimation, my approach is adequate, and my guess is it would be correct to 10%.

Will it stand up in court?

I knew legal holes which the other side could shoot into my argument apart from the accuracy problem, which is a common first line of criticism. How did I know whether the rod was straight in the first place? I didn't. Of course, you ask the owner, he will swear it was straight because otherwise it would be admission of another deficiency on his part. In any case, it would still be a conjecture on my part, an 'expert opinion', to be argued endlessly as a stroke of genius by my lawyer, and as a crude flaw by the other side.

Actually, I did not have a chance to find out. The investigation never became a case, but was settled out of court – as happens in most cases that do not require mandatory case filing.

(b) Underpass width:

Another time, I had to find the distance between two very critical lines on a surface in a fall accident, and give my expert opinion on how it compared with distances between similar lines in various countries.

The Internet had the information only for a few of them. I had to rely on pictures captured from the Internet to make my argument. I resorted to computer graphics.

To demonstrate the procedure I adopted, let me take the example of finding the width u of a London underpass in Fig. 3.17 (left) – nothing to do with my case, which I cannot talk about, yet.

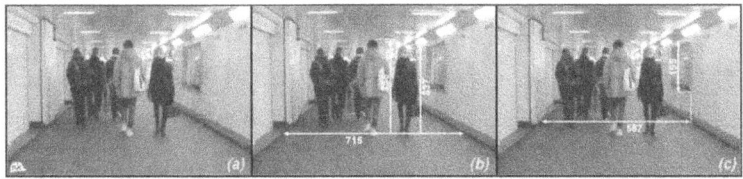

Fig. 3.17 Estimation of the underpass width from a photo

To find u from the photo, I can try to relate it to the height of the image of the people walking. Picking the nearest man and woman pair, I measure their heights as 355 and 337 pixels, and width of the underpass at their feet as 715 pixels (Fig. 3.17, middle). Then u will be [(man's height/355) or (woman's height/337)] times 715.

I have now fallen from the frying pan into the fire – I don't know these two people!

However, the Internet has lots of statistics on heights of people in various countries. For England, BBC gives men's and women's average heights as 1.75 m and 1.62 m, [3.17].

From this, my estimate of u = 1.75×715/355, or = 1.62×715/337, i.e., 3.52 m or 3.44 m; quite close, giving an average of 3.48 m, maybe round off to 3.5 m. That is the best I can do for now.

Suppose I know that the height of the notice board on the right wall is exactly 1.1 m (Fig. 17, right).

Then, finding the height of the board and the width u of the underpass in the picture in pixels at the same view distance as 195 and 587, I get u = 1.1×587/195 = 3.31 m.

So my estimate from people height is in error by 100×(3.48–3.31)/[(3.48+3.31)/2), i.e. 5%.

Not bad! But then, first I was lucky to get 'average' Britishers; then I was able to bracket the error only because I had the exact dimension of something in the same picture as the people. So without these, the only thing I can do is to minimise my potential error in various ways.

In general, an investigator will have to allow for the following variables. If he does not, the other side will tear his findings to pieces.

- The variation of heights from the average for the majority of people in a set may be ±5%, a range of 10%.

- The difference between average male and female heights runs between 8 and 10%. So, if you guess the gender wrongly (as is quite possible in these days of unisex dress and hairstyles, especially from the back) your results are off by another 10%.

- The person(s) in the picture may be tourists from another country with average heights much taller (Danish, 184 cm/171 cm, about 5% more) or much shorter (Philippines, 162 cm/150 cm, about 8% less) than the subject city's average height? These would introduce another variation of another 13%.

Add it all up, and we end up with a variation of 33%, in the worst case scenario of the investigator mistaking the shortest lady from the Philippines for the tallest man from Denmark! It is really not so bad in practice.

To minimise the error, I usually look for a picture with a group, and pick an 'average' person whose gender is fairly clear. Then I allow for a height variation of ±5% within a homogeneous group. If the person is obviously from a particular region, I make broad allowances for the height variation for that region also, such as Europeans versus Asians.

Ultimately, I aim at an error potential of 10%, which is quite good when you are dealing with a large number of cases. Of course, such precision cannot be guaranteed when a single dimension in a single picture has to be estimated.

In the particular case analysed, I could have given my estimate (without the board height) as 3.48±0.35 m. Then the worst error from the correct 3.31 m would have been 100×(3.48+0.35−3.31)/3.48, i.e. 16%, I would say still not too bad.

Beyond all these are the complexities of the graphics perspective itself.

Best estimates are obtained when the camera is pointing directly along the centre line of whatever you wish to estimate, and held about waist level of people, as in the case presented. But the camera may be to one side or the other, or higher than people level, or the vista may be tilted or curved.

You can't give up – you must still put your best foot forward!

I try to calibrate –use a picture where I know the dimension sought or some other length, apply the pixel estimate method, and adjust my computed value for any difference for that and similar pictures – the factor would take into account camera focal length, position, etc.

If there is a noticeable horizontal and/or vertical angle between the camera axis and the line perpendicular to the measurement sought, I try to apply corrections involving the angle.

I also try to find pictures with more than one person – not in the same line as I did in this example – from whom I can determine the desired dimension, and take the average for different persons. (In Fig. 3.17, I had two or three more chances.) If I get different pictures of the same scene, I take results from different scenes and average again, improving the estimates each time.

When I use this technique, I should be prepared to satisfy the other side and the court. First, for the record, I must explain the graphics method I use, because it may not be a routine procedure. I must list the approximations I make and the resulting range of error.

I may be asked how I know that the original pictures from the web had not been modified graphically to start with. I don't, but I may explain that these were record shots and not promotional ones for the particular dimension in question – for which again many and multiple sources are the best bet.

Will it survive in court? No comment – it depends on how it is expected to be used.

In a similar approach, I have used trajectories of moving objects by extrapolation of motion paths in individual frames of videos, to guesstimate starting/ending points of moving objects.

3.9 ANIMATIONS AND SIMULATIONS

Animation and simulation are at the top of the list of computer applications in forensic engineering. If you can do it well, you have a powerful forensic tool.

Of course, even if you cannot do it yourself, it is as well if you have the ideas and can get someone else to do it well for you, as clients will be happy to bear the extra charges for additional information you produce.

3.9.1 Difference between animation and simulation

Both animation and simulation involve graphic images in motion, but there is a difference, even separate legal definitions, in many Western courts.

Bow Tie Law's Blog, [3.18] differentiates between them as follows:

"A computer animation has the following characteristics:

- *Moving pictures not intended to simulate an event;*
- *Authenticated by a sponsoring witness with personal knowledge of the content of the animation;*
- *Showing that it fairly and adequately portrays the facts; and,*
- *Helps illustrate the testimony.*

"A computer simulation has the following characteristics:

- *Scientific evidence;*
- *Generally detailed and realistic recreated computer image of the event that can be manipulated; and,*
- *Can be portrayed from different angles or from the viewpoints of different witnesses. "*

My work on the falling worker in Fig. 3.12 is an animation and not a simulation, because it depicts parametric results with assumed data, and did not simulate any specific fall. If I add motion from dynamics equations applied to specific accident data, it may qualify as a simulation.

One must have permission from the court even to show any slide or video of factual matters, and the same applies all the more seriously to imaginary visuals. Animation/simulation must be:

(i) Authentic,

(ii) Relevant;

(iii) A fair and accurate representation of the evidence to which it relates; and,

(iv) Of probative value that is not substantially outweighed by the danger of unfair prejudice.

At first one would go gaga at how many tricks could be done with animation tools.

But it is not worth getting too cute with them, because smart lawyers and judges can see through them. Soon the visuals could begin to have a negative effect, like some slide presentations do where the colours and the slide changes look (and sound) like Diwali

(Indian Festival of Lights) fireworks.

3.9.2   Animation

A lot of animation is used in traffic engineering cases, because that is the easiest way the audience can understand the position of various vehicles involved in the accident before, during and after the event.

Before computers became so powerful, and even now in the smaller (low budget) cases, lawyers and judges manage with coloured blocks of appropriate sizes being moved around to demonstrate different views of what each side wants to prove happened.

But today, almost all traffic cases come with nice video presentations of the different scenarios, often supported by car-mounted or street-corner videos. While judges are not to be influenced by the action and colour, it is difficult not to be impressed by a well prepared video!

With my crush on computer (or any!) graphics, it has been natural for me to try animation whenever and wherever I could. Here I give one professional example from a litigation consultant on the internet, and a couple of my own minor examples (which I can talk about) – nothing most others cannot do:

(a) Site accident animation and physical model:

Figure 3.18 shows stills from a video by Z-axis Litigation, [3.19], of a construction worker who is pulling some planks to make a work platform and in the process falls off the work area to his death. Top row in Fig. 3.18 is 3D rendering of the scene, and bottom row shows parts of the animation of the event and a physical model.

His survivors claimed that the employers did not provide fall prevention safeguards, but the forensic evidence, bolstered by the animation video and one-third size physical model, proved that the worker had failed to anchor the safety harness he was wearing to the safety line which had been within his arm's easy reach. His family did not get any compensation.

Fig. 3.18 Construction accident animation and physical model

(b) Animation of rebar support problems and solutions:

In one presentation about a case, I used animation to show how one rebar rod would slide around another if laid on top of each other, as shown in Fig. 3.19 (left). It looks clear enough in a still picture, but when displayed actually sliding, it is a dramatic demonstration.

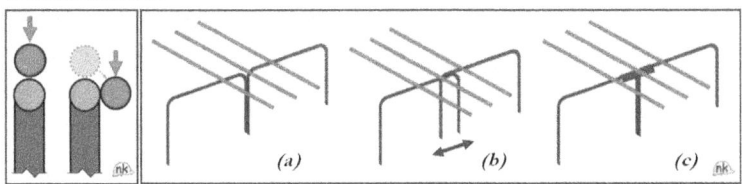

Fig. 3.19 Animation of rebar support problems and solutions

In the same case, the contractor had argued that he was forced to change the in-plan positioning of support frames for rebars as shown in Fig. 3.19, right, (a) to (b) so as to avoid lack of support where the bars were atop the bent corners of the frame as in (a).

Irrespective of whether this aggravated the collapse potential or not, the point was that this was a deviation from the design, and I pointed out that it would have been simpler, cheaper (because of no wastage of overlap in the contractor's solution), and better adherence to the design if he had put a cap or even tied a piece of rod over the bend gaps as in Fig. 3.19, right, (c).

For greater effect, I animated the two frames sliding from (a) to (b) to overlap each other and support the middle rebar, and then sliding back from (b) to (c) with a cap to cover the gap. This eliminated any possibility that everybody except the contractor and I would be mystified.

3.9.3 Simulation

Computer simulation, when permitted and done well, can be very effective. It does not demand search for and acquisition of parts to build your experiments, waiting for materials, technicians, or fabrication.

And if it does not work like you expected, you can simply change one or a few numbers in the input and re-do the analysis almost immediately.

The best part of it is that it does not hurt anybody or damage anything if the product or structure fails.

You can recreate described conditions, or present alternative configurations much easier in a computer than building a physical prototype as was the practice before computer graphics. The colour, texture, shadows, almost everything that can describe a real object and its surroundings may be represented. This saves a lot of time and effort in court testimony.

You can play around to your heart's content and try out even stupid-sounding ideas without anybody else being the wiser about your foolishness!

Of course, you need to confirm, or calibrate your computer model with a few prototype or model tests. Otherwise it would be just another animation.

Although personally I have simulated many events and processes both in the lab and through finite element method in my structures research and consulting, I have not had the opportunity to produce a simulation for expert forensic testimony thus far. I will present a few examples from public domain.

(a) Precast Panel Fall on Worker:

Any amount of verbal description accompanied by still

photographs will not bring home the trauma of an actual accident like a video can, as indicated by the four stills shown in Fig. 3.20 from the one-minute long 3D-video simulation of a crane operator controlling the lifting of a precast wall panel, and the panel breaking away from the crane hooks and falling towards the operator. He tries to escape, but the wall catches him on the back of the right leg. [3.20].

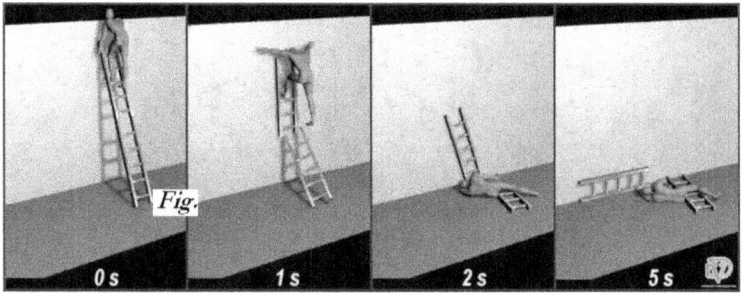

Fig. 3.20 Man falling off of ladder (Partial frames only)

The video was specially made for presentation in a court case. In this simulation, all the objects are realistically digitised to scale, and the motion is implemented to represent scientific principles. The view is continuously displayed from different positions so that the forensic engineer can interactively show the breaking away of the anchors and the fall of the wall.

(b) World Trade Center Collapse:

The World Trade Center (WTC) terrorist attack on 11 September 2001 in New York is probably the world's most investigated disaster of modern times, taking as it did a human toll of nearly 3000 innocent lives at one time. Apart from scientific curiosity, government imperatives and the human dimensions, there was also the wild charge that the collapse was 'engineered' by the Government for political ends that forced such extensive forensic investigations.

If we can look beyond the human tragedy, the numerous finite element analyses of the buildings and events involved make a beautiful application of computer technology in the representation of reality and pursuit of truth.

A most impressive one is the Purdue University version by FEA of the WTC North Tower attack, [3.21]. This is mostly simulation in the sense that the structure and interacting elements and processes are all scientifically accurate.

But some animation has been added on to show effects such as flames and smoke. Figure 3.21 displays the major features of the finite element modelling.

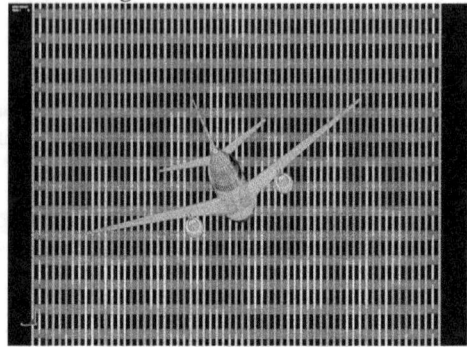

Fig. 3.21 Purdue University WTC disaster finite element model

Figure 3.22 shows screenshots from the 5 minute video simulating the crash and animating some of its effects, as follows:

1. Title and acknowledgement
2-3. Google Earth images, plane hitting, fire starting
4. Plane entering tower
5. Plane almost fully entered, in a fraction of a second
6-8. Interior view, plane entering, crashing through
9. Plane pieces exiting from tower
10-12. Interior view, fire starting and spreading
13. Exterior view, fire spreading
14-15. Only tower columns shown, plane hitting and slicing columns
16. Rendering of dust and glass pieces

The video is amazingly realistic, slowed down considerably to depict the simulated crash.

What came out of such forensics was the fact that both the North and South towers survived so long despite the loss of many supporting columns, allowing tens of thousands of occupants to rush down to safety, because the original designers had left quite a bit of redundancy in the structure in the form of steel window mullions which took over when the core columns failed.

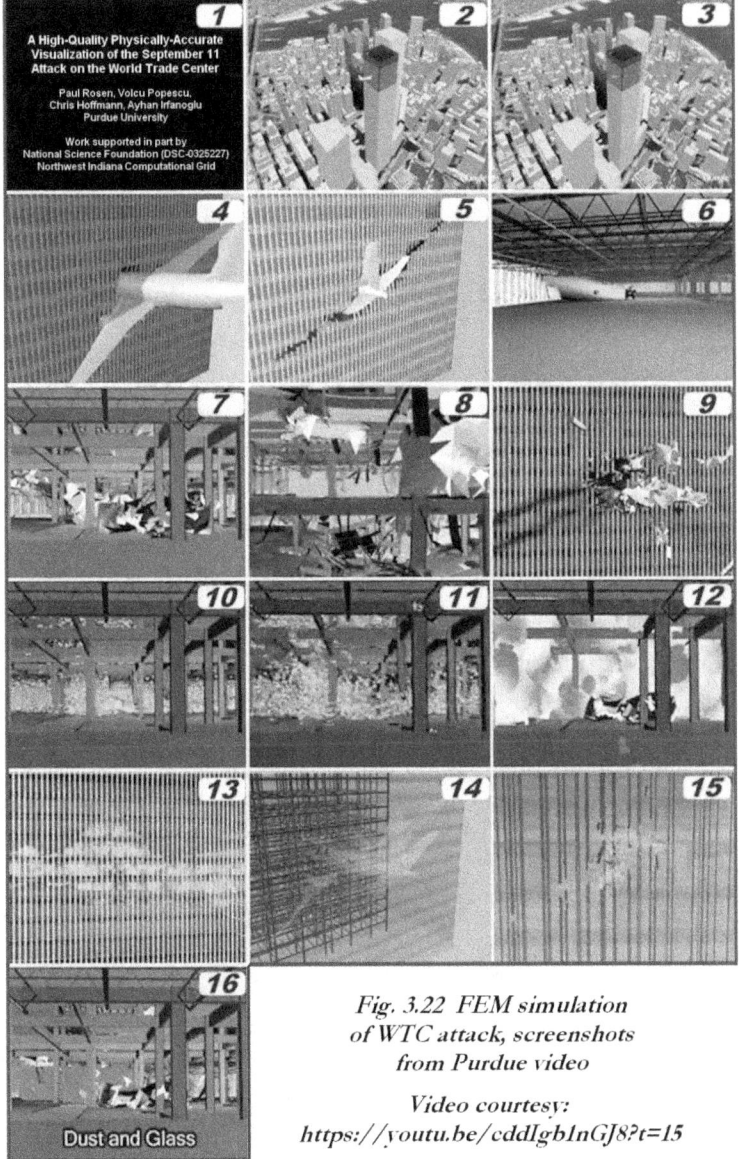

Fig. 3.22 FEM simulation of WTC attack, screenshots from Purdue video

Video courtesy:
https://youtu.be/cddIgb1nGJ8?t=15

Main lessons learnt were to increase the width of stairs to permit faster evacuation, and better fire protection.

3.10 CONCLUSION

Computers are here to stay. We need them in forensic engineering as much as in everything else, but we must be careful with them, particularly with powerful software which can make our work so much easier if used right, but which can land us in deep trouble if we misuse them.

3.11 REFERENCES

3.1. _____, *Failure Analysis of Minneapolis I-35W Bridge Gusset Plates*, Simulia. Retrieved July 2013: http://www.3ds.com/fileadmin/PRODUCTS/SIMULIA/PDF/tech-briefs/Arch-Failure-Analysis-of-Minneapolis-I-35W-Bridge-09.pdf

3.2. LaMalva, K.J., *Complete Report on Failure Analysis of World Trade Center 5, 7p*. Retrieved July 2013: http://www.engineeringcivil.com/complete-report-on-failure-analysis-of-world-trade-center-5.html

3.3. Krishnamurthy, N., "Investigative Methods in Forensic Civil Engineering", *Proceedings of the Conference and Exhibition on Forensic Civil Engineering*, ACCE, 23-24 August 2013, Bangalore, India.

3.4. _____, *Pentium FDIV bug*, Wikipedia. Retrieved July 2013: http://en.wikipedia.org/wiki/Pentium_FDIV_bug

3.5. _____, "Error in proprietary design program", Report cros349, *CROSS Newsletter*, No. 31 – July 2013. Retrieved July 2013: http://www.structural-safety.org/view-report/cross416/

3.6. Delatte, N.J.,M. and K. L. Rens, "Forensics and Case Studies in Civil Engineering Education – State of the Art, *ASCE Journal of Performance of Constructed Facilities*, Vol. 16, No. 3, Aug. 2002, p. 98-109.

3.7. Krishnamurthy, N., *Three-Dimensional Finite Element Analysis of*

Thick-Walled Vessel-Nozzle Junctions with Curved Transition, ORNL-TM-3315, Oak Ridge National Laboratory, U.S. Atomic Energy Commission, July 1971.

3.8. Bell, G.R. and A.A. Liepins, "More Misapplications of the Finite Element Method," *Forensic Engineering: Proceedings of the First Congress,* (K.L. Rens, Ed.) ASCE, Minneapolis, USA, 1997, pp. 258-267.

3.9. Krishnamurthy, N., "Investigative Methods in Forensic Civil Engineering", *Proceedings of the Conference and Exhibition on Forensic Civil Engineering,*23-24 August 2013, Bangalore, India.

3.10. Johnson, R.G., *Hartford Civic Center,* 2009,. Retrieved July 2013: https://failures.wikispaces.com/Hartford+Civic+Center+(Johnson)

3.11. _____, *The sinking of the Sleipner-A offshore platform.* Retrieved July 2013: http://ta.twi.tudelft.nl/users/vuik/wi211/disasters.html

3.12. Emkin, Leroy Z., *Comparison Of Static Analysis Results Based On Different Models Of A 67 Story Commercial Building,* 2002, 113 p. Retrieved July 2013: http://www.gtstrudl.gatech.edu/gtstrudl_new/GTSTRUDL_Users_Group_Newsletter/UsersGroup/2002/Emkin_2002.pdf

3.13. Krishnamurthy, N., "Safety in High-Rise Design and Construction", *Seminar pre-print, published in ' Build Tech – 2006 Souvenir of International Seminar on High Rise Structures',* by Builders' Association of India, Mysore Centre, Dec. 2006, p. 19-34. Retrievable from: www.profkrishna.com/ProfK-Publications/Publications.htm

3.14. _____, "Chapter 5 – Causes of the Collapse and Findings", *Report of the Committee of Inquiry into the Incident at the MRT Circle Line Worksite that led to the Collapse of Nicoll Highway on 20 April 2004,* Figures 5.2, 5.3.

3.15. Krishnamurthy, N., "Worker fall from mobile scaffold", *Int. J. Forensic Engineering,* Vol. 1, No. 1, 2012, p. 21-46.

3.16. Krishnamurthy, N., *Introduction to Computer Graphics.* Tata-McGraw-Hill Publishing Co. Ltd., New Delhi, India, 2002, 343p.

3.17. _____, *Statistics reveal Britain's 'Mr and Mrs Average',* BBC, http://www.bbc.co.uk/news/uk-11534042

3.18. _____, *Computer Animations vs. Simulations: What is the Difference?* Retrieved Jul 2013: http://bowtielaw.wordpress.com/2009/02/20/computer-animations-vs-simulations-what-is-the-difference/

3.19. _____, *Construction Site Accident Animation and Physical Model,* Z-Axis Litigation. Retrieved July 2013: http://www.youtube.com/watch?v=c-KySsKVp20&feature=c4-overview&list=UUJxyMA8mKyrZQoDTh_jFjAA

3.20. _____, *Construction Accident Recreation.* Retrieved July 2013: http://www.youtube.com/watch?v=_XSJTq4n6ug

3.21. _____, *Purdue creates scientifically based animation of 9/11 attack,* Purdue University News, 12 June 2007, 4p. Retrieved July 2013: http://www.purdue.edu/uns/x/2007a/070612HoffmannWTC.html

---

† This is a reprint of the author's paper presented at the Forensic Civil Engineering Conference and Exhibition held at Bangalore, India, on 23-24 Aug. 2013, by the Association of Consulting Civil Engineers (India), reprinted with permission.

4

EXPERIENCES OF AN EXPERT WITNESS†

4.1 INTRODUCTION

The word 'forensic' originates from 'forum' which in use has come to mean legal assemblies.

In Civil law the disputes are between 'plaintiff' and 'defendant' individuals or organizations. Criminal law deals with crime and legal punishment where the Government is always the 'prosecution', and the charged person or organisation is the defendant. It may also happen that a private plaintiff sues a statutory body which then becomes the defendant.

A forensic engineer will have the opportunity, privilege and challenge of appearing in Court to testify on behalf of a client or an employer (both referred to as 'client' in the rest of the paper) for Plaintiff (or Prosecution) or Defence, or even a third party involved in a case.

Not all accident investigations end up in Court. Many Civil cases get settled out of Court when the accused party finds that proof against it is too strong, and prefers to settle the issue with the plaintiff for a mutually acceptable compensation rather than take it to Court and after a long time and much expense lose the case. Mediation and arbitration are other options.

I have appeared as expert witness on behalf of the Government, companies, and injured victims. But I am no lawyer. What little I know by testifying in Court has taught me how complicated law is or can get. I will therefore not discuss legal procedures, and defer to lawyers to clarify legal points. I will only comment on my ventures into the legal systems with which I have experience as a forensic engineer and expert witness.

Any error in procedural details may be attributed to my ignorance, or fading memory – I do not wish to be held legally responsible!

While a 'lawyer' is anybody who has passed a law examination, an 'attorney' is one who is licensed to practice law in courts. However, I shall use both terms interchangeably in this paper. I will use following abbreviations to save repetition of same phrases many times:

      EW – Expert Witness, I, you

      OC – Opposing Counsel, the other side

      WC – Witness's Counsel, my side, your side

Use of male pronoun or other reference would automatically cover the female equivalent except when either one is implied by context.

Due to legal constraints, I will not be giving actual names, dates, places, or identification.

4.2 WHO IS AN EXPERT WITNESS?

The West recognises a difference between an "expert consultant" and an "expert witness". A consultant (or adviser) may be someone whom a client believes to be especially knowledgeable to assist with certain technical aspects of a problem. When the expert is retained to testify in Court based on his credentials and findings, he becomes an EW.

4.2.1 Definitions of Expert Witness

Among the many definitions of an EW that exist, I prefer to accept Ratay's choices, [4.1]: *(i)* *"A non-biased professional who provides testimony based upon his/her technical expertise." (ii) "A person who, by reasons of education or special training, possesses knowledge of some particular subject area in greater depth than the public at large."*

However, only the courts decide if one is allowed to testify as an EW.

According to US Federal Rules of Evidence, an expert may be allowed to testify under following conditions:

- Scientific, technical, or other specialized knowledge will assist the Court to understand the evidence or to determine a fact in issue;
- Witness can be qualified as an expert by knowledge, skill, experience, training, or education;
- Testimony is based on sufficient facts or data and is the product of reliable principles or methods;
- Witness has applied the principles and methods reliably to the facts of the case; and,
- Opinion is sufficiently relevant to the facts of the case.

ASCE's guidelines on forensic engineering, [4.2] defines four criteria for the admissibility of what the expert testifies:

1. Peer review and publications,
2. Rate of error,
3. Empirical testability, and,
4. General acceptance within the scientific community.

4.2.2 Qualifications for an Expert Witness

Both the courts and the industry recognise that an EW must

possess sufficient expertise and mastery over his claimed area of speciality by demonstrating following five key attributes:

1. Education,
2. Training,
3. Experience,
4. Skill (evaluated as accuracy *plus* efficiency *plus* timeliness), and,
5. Knowledge.

These overlap the qualifications for being a forensic engineer, which I have discussed in another paper, [4.3].

The special attributes to be an EW would be the ability and willingness to face intense questioning under stress and respond with clarity and confidence. The attributes listed here may overcome any deficiencies in other factors that are considered basic to a forensic engineer, such as not being a P.E. I am not a P.E., but I am accepted as an EW.

4.3 BECOMING AN EXPERT WITNESS

4.3.1 Getting started

Talk to an EW. Attend one or two Court sessions when EW is on the stand. Understand all the legal nuances of being an EW. Then if you still want to become an EW, get going.

- Advertisements must be carefully drafted with legal advice. What you advertise may be brought up in Court to your disadvantage. However, personal advertising has not been found to be too effective in getting forensic business.

- Join a forensic engineering company; most advertise their services.

- Contact legal firms known for accident litigation. They get accident litigation business first and foremost; lawyers seek out EW. Insurance companies are a close second.

- Inform your local professional society of your interest. They get queries for experts. Institution of Engineers (Singapore) has an 'Expert Panel' for members to apply.

- If you are good in design and safety areas and already a consultant, clients will come to you, or somebody will refer a client to you. University faculty are in demand when an accident investigation becomes necessary, because of their publications and seminar lectures. Their ex-students will be good contacts. Such referrals are the best.

- If you speak at conferences/seminars, mention your interest in accident investigation.

In sum, the more you present your ideas to various groups, the more people will remember them and contact you.

The first case will be your most difficult one. After that, demand will go up (especially if you do well) and soon you will have enough to pick and choose.

Personal Experience – 1:

In USA of the 1960s, it was my specialising in the (then new) area of Matrix and Finite Element Analysis that gave me a boost to my research and consultancy. Soon I started getting requests to become an EW. But, I decided not to become an EW myself (mainly because of the high professional liability) but to act as consultant for accident investigators and EWs with my back-up analysis and testing. That was how I learnt the ropes of being an EW.

In Singapore, even before I could think of actively entering forensic engineering, I received enquiries and assignments by referrals. My courses, seminars, workshops and publications on workplace safety and risk management also helped my forensic engineering.

(1-End)

4.3.2   Which cases to accept

A forensic engineer may choose to remain an expert consultant and not an EW (as I was in the USA).

But you must tell your client up front, and not half way through, that you don't want to be an EW. On the other hand, you may be directly requested to be an EW.

Anyway, the acceptance of an assignment should be a carefully considered decision to accept to be an EW, and not a casual one.

All Court appearances are stressful in one way or another, to a smaller or greater degree. Lawyers are dependent on the EWs they select, and are also subject to stress from them and their performance in the case. So your wrong decision may hurt more than just you.

Shuirman and Slossom, [4.4] give a checklist to decide how to accept an assignment:

1. Be certain there are no conflicts of interest with any of the parties connected with the lawsuit, directly or indirectly.
2. Obtain as much background information on the accident or failure as possible, mindful that such information may be biased or lack pertinent data
3. Attempt to separate facts from opinions so as to be able to form an objective early picture of the main issues.
4. Inquire as to the status of the case and its tentative schedule to determine whether there is sufficient time for a thorough investigation.
5. Make certain the matter is within the appropriate areas of expertise for you or your firm.
6. Discuss fee schedules and determine when and by whom payments will be made.
7. Check out the reputation of the lawyer or law firm if it is not already known.

Personal Experience – 2:

Probably because of my age (82) and current indifference to a career as such, nowadays I take up only those cases that satisfy two personal criteria:

- It must be challenging and not routine (which <u>anybody</u> can do); and/or,
- It must be something through which I can constructively benefit some individual who I believe could have been better

protected against loss or injury, or benefit some industry or society in general, in a way that would improve the safety culture.

(2-End)

4.4 ROLE AND RESPONSIBILITY OF AN EXPERT WITNESS

4.4.1 Obligations of an Expert Witness

The forensic engineer/expert witness has dual obligations, [4.1]:

- To the Court, assist in understanding the complicated technical subjects that are not within the knowledge of the average person.

- To the Client, assist the attorney with the technical information he/she needs to develop the case, provide expert opinion, and give expert testimony, if called upon.

The Inter-professional Council on Environmental Design (ICED)/Association of Soil and Foundation Engineers (ASFE) has a very good list of obligatory directives for EWs, [4.5]:

- Avoid conflicts of interest and the appearance of conflicts of interest.

- Undertake an engagement only when qualified to do so, and when you can rely upon other qualified parties for assistance in matters beyond your area of expertise.

- Consider other practitioners' opinions relative to principles associated with the matter.

- Obtain available factual information relative to events in question to minimize reliance on assumptions, and be prepared to explain any assumptions to the Court.

- Evaluate reasonable explanations of causes and effects.

- Assure integrity of tests and investigations conducted as part of the expert's services.

- Testify about professional standards of care only with knowledge of those standards which prevailed at the time in question, based upon reasonable inquiry.
- Use only those illustrative devices or presentations which simplify or clarify an issue.
- Maintain custody and control over whatever materials are entrusted to your care.
- Respect confidentiality about the assignment.
- Refuse or terminate involvement in the engagement when fee is used in an attempt to compromise your judgment.
- Refuse or terminate involvement when you are not permitted to perform investigation you believe is necessary to render an opinion with a reasonable degree of certainty.
- Maintain a professional demeanour and be dispassionate at all times.

4.4.2   Role of Expert Witness

Ideally, the expert discharges his obligations to both the Court and to the client equally – but justice's demand for truth may be in conflict with client's need for helpful testimony.

You must remember that you are first an expert and only second a witness for one side or the other in the litigation.

Before you accept the assignment you must be fairly clear if the aspects of the case you will be analysing and presenting will be (mostly) for or against the potential client.

If you go into a case blind, you will be either pressured to limit your testimony to only the supporting points (if not actually 'massage' them in the client's favour!) or you will be summarily dropped from the case, hopefully after payment of some fee.

In the words of ASCE, [4.2] *'Forensic engineers do not win or lose trials; lawyers do. The Forensic engineer, therefore, has no track record of success or failure in litigation. The Forensic engineer's role at trial is to present opinions in a believable and credible way. That is the ultimate determination of success as an expert witness.'*

It is hard for EW to remember he is not 'working for' the client who pays him.

It is harder for the client who is paying for EW's services to accept EW is not looking out solely for him.

Clients are jubilant when they win and want to pay you more, but they are bitter and recriminatory when they lose. Lawyers do tell the EW and the client that the EW's role is to bring in technical clarity to the case and not take sides. But they too have a stake in the EW because usually they are the ones who pick and recommend the EW to the client.

4.4.3   Selective support by Expert Witness

If you cannot or do not want to handle all aspects of a case, you should tell the client the strong and weak points of the case, and detail which portions you will support and which not.

- But this is not a complete solution, because the opposing side will be watching for areas that you do not touch, and pounce upon them demanding your opinion, so that anything against your side and/or supporting their side may be documented by it.

  So you will have to be careful how you answer questions involving your judgement.

- The OC may ask, *"Do you agree that your client was wrong?"* You may respond with: *"The outcome might be wrong, but I do not know the circumstances of who made the decision, or how and why."* A simple *"No"* may lead to further questions.

- Meanwhile, your lawyer can (and should) object to the OC 'badgering' you, and the judge may get you off the hook.

  As a consultant to your client, you have a professional responsibility to assist him – technically, not legally – to the best of your ability.

  However, your responsibility does not extend to carrying out your client's wishes to slant your testimony in his favour contradictory to facts or good practice.

Your duty to truth and society transcends your loyalty to your client.

If it gets bad after you have started, you may withdraw, but it will hurt your reputation.

The lawyer's role is clearer, charged with representing the client, managing his litigation, and winning the case for him. He selects an EW who he thinks will help the case, and not just find truth.

He will seek and use only information that will help their case. So, he will manage and use EW's testimony to maximum advantage, without of course violating facts or truths.

The EW in Court should not volunteer all the information he knows on the case, but should only respond to the lawyer's questions from both sides.

- However, if OC directly asks your opinion on a critical matter and you know the answer, you must give it (limiting it to the absolute minimum) damaging though it might be to your client's case.

Otherwise you run the risk of being branded as not being so expert after all, or charged with being unethical in your profession.

The responsibility of the EW is to investigate the failure or accident as thoroughly and fully as possible. But the form, extent, and content of the disclosure are the client and lawyer's choice, as long as EW's representation of them is truthful.

OC may well ask: *"Are you being paid for your testimony?"* Although it is quite fair that you are paid for serving as EW, ASCE, [4.2] recommends that you answer, *"I am being paid for my time"* [and I may add, *"my expertise"*] *"and not for my testimony."*

4.4.4   Adversarial relationship

Although the dispute is between two litigating parties and <u>not</u> between the two experts, each EW, however senior he may be, should be prepared for confrontation from the opposing side.

Courts are adversarial in nature. The two sides are expected to fight it out in Court, with judge as referee. So OC will hit hard on the EW at every opportunity. Counsels get a lot of leeway in their examination so that they may ferret out truth, which after all is what judges also want. Judges may often overrule WC's objection to OC's aggressive line of questioning.

- The other side will certainly have expert opinions to support their side and quash your side. It is your responsibility to compare both sets of ideas and clarify the issues.

- Even in science-based engineering, there could be more than one way to solve a problem, more than one type of curve to fit a set of data, more than one cause for a failure, and most important truth of all, more than one trigger for an accident.

    So don't be surprised or upset if your pet theories to explain your client's side of an accident are shot down or at least countered by an equally hot theory from the other side. You will be smart to think of all possible scenarios so that you won't be caught napping.

- When an alternative conclusion is offered by other side, accept it in principle unless you can counter it with valid arguments. When new facts are adduced by them, accept them gracefully and attempt to modify your own findings to fit the new information.

Despite this tension, you must maintain a professional demeanour and courteous interface with opposing lawyers and experts, and not try to 'score' points for your side at all costs.

Sometimes, in spite of your best efforts, all your labours might seem to have been in vain.

Personal Experience – 3:

In the case of a formwork failure, I came up with three scenarios for the prosecution and the defence came up with a fourth. But on the last day of testimony, a witness presented evidence which produced a fifth feasible scenario. The judge simply threw out the case as the prosecution could not prove their case 'beyond reasonable doubt' – and this, without either side being able to prove the other side wrong without their own testimony!

(3-End)

4.4.5   Relationship to the legal system

The EW is obligated to the lawyers on both sides and the judge for certain acts:

- The EW may generally be required to explain and simplify technical points to the judge and legal teams as well as to the media and through them to the public.
- He should participate in strategy meetings with his lawyers.
- He must participate in discovery disclosures (i.e. preliminary meetings of the legal teams), technical requests, discussion of cross-examinations by opposing experts etc.
- He should satisfy the Court of his credentials and answer all questions on them.
- He should abide by the 'Do-s' and 'Don't-s' of courtroom and testimony protocol proffered by the lawyers.

In all the preceding, he must not be influenced to re-align his comments in any way just to benefit their side, but be objective, and not stray from his convictions.

EW is in the unique position of not being part of the dispute, and only contributing his expertise on behalf of his client, and to that extent, he can be an instrument of revealing the truth and promoting progress in his speciality for the benefit of all concerned.

- If both sides agree, both experts can meet and discuss their differences, and if the problem will admit a consensus solution, present it to both sides and aim at mediation, and resolve precise measures to repair the damage and settle the compensation claims in a fair manner. This is good mediation. But first, both clients must like the idea!
- Out of the confrontation between opposing EWs, the expectation is that the judge can make a final decision based on

the strength of the comparative arguments.

4.5 LEGAL PROCEDURES

4.5.1 Steps in a non-jury trial

In USA and UK most trials are by jury, but in Singapore and India the jury system has been abolished. The general procedure for non-jury trials is as follows:

1. Report of forensic engineer's investigation, his recent relevant CV, and supporting documents legalised with an affidavit will be submitted to the Court and the other side as 'Discovery'. The forensic engineer may be called for a discussion with both counsel and a deposition made of his responses. Based on the reports and discussion, decision on whether or not to call the forensic engineer as EW will be made.

2. Once testimony starts, the Prosecutor/Plaintiff lawyer calls his witnesses – this initial testimony is called 'Direct' testimony, or 'Examination-in-Chief';

3. Defence lawyer may cross-examine the witnesses;

4. Prosecutor/Plaintiff lawyer may do further questioning to clarify certain answers brought up during cross-examination – this is called 're-direct' or 're-examination';

5. Sometimes defence may again cross-examine the witness – this is called 're-cross';

6. Prosecutor closes the Government's case, or Plaintiff rests;

7. Defence may call witnesses, and Prosecutor/Plaintiff lawyer may cross-examine them;

8. Defence rests;

9. If the Prosecutor/Plaintiff presents "rebuttal" witnesses/ evidence to challenge evidence presented by the Defence during their phase of the trial, then there is another round of examinations before the Prosecutor/Plaintiff rests;

10. Prosecutor or plaintiff's lawyer presents a closing argument to the judge;

11. Defence lawyer presents a closing argument to the judge; and finally,

12. Judge returns a verdict, after an interval of time.

4.5.2 Legal role of an Expert Witness

If a deposition is taken from the forensic engineer, both parties will get an opportunity to fully explore opinions that he will offer at the trial, as well as to determine his credibility as an EW. Deposition can be lengthy and go into great detail, and as such can be very stressful to the forensic engineer. Often, depositions may be by report and affidavit from a consultant.

Trial testimony on the other hand generally will focus on key issues, aimed at searching for the truth. This will however place the EW in the spotlight where every word and every gesture will count for or against him and his side.

Communication between experts and lawyers enjoy attorney-client privilege.

But the forensic expert must be careful regarding the material he retains with him on the case because all materials hand-written, typed, electronic (including e-mails and blogs), audio- and video-tapes, photographs, sketches, drawings, doodles, notes, diaries, memos, letters, working drafts as well as final reports, etc. can be demanded by the opposing side through the Court.

Bad news of a provable mistake by the client – after thorough confirmation by the expert – must be communicated to the client and the lawyers as promptly and clearly as good news. If some error is not informed and opposing side finds it out and brings it out, it will reflect badly on the EW's credibility and ethics, apart from damaging his client.

It is wise not to put bad news in writing of any form. In certain privileged relationships, these may not have to be disclosed. It would be best to be guided by counsel in this matter.

In case a site visit or examination of designs or drawings shows a highly hazardous situation which may escalate into a catastrophe involving human injury, environmental damage, or great property loss, the forensic engineer must first inform the employer/client and if no immediate response is forthcoming, then inform the appropriate public official about the danger, without revealing unnecessary incriminating information.

The judgement is mostly based on the evidence presented by the two sides. Most law is argued on the basis of precedents, that is, on past cases with similar features. Lawyers on both sides scramble to find cases where the judgement has been delivered under similar circumstances in their particular side's favour, and it is for the judge to sort out the validity and strength of the two sets of arguments, utilising additional precedents of his own.

That a judgement itself can be arguable if not outright wrong is borne out by the fact that many times the judgement of a lower Court is reversed or annulled by the higher Court.

4.5.3   Witness Counsel's support to Expert Witness

Lawyers confer with EW many times and at length with a three-fold purpose, to:

(i)   Brief EW on Court protocol and coach EW on their line of questioning;

(ii) Be briefed on the technicalities of EW's work and learn about fruitful lines of questioning to give EW full scope to explain or clarify; and,

(iii) Understand other side's technicalities, and frame questions to attack their weak spots.

Of these, the first function is very critical to the EW. Lawyers are usually smarter and better-read than most lay persons. They are more aggressive and better communicators. The EW better follow counsel's advice on Court behaviour. They may even have tidbits about the particular judge you will be facing, what he tends to overlook and what he gets upset about.

If you are new to the game, lawyers will tell you how to dress, how to sit, stand, move your hands, keep your face from smirking or frowning, not hem or haw, not get upset, not try to score debate points, not interrupt the questioner – and not answer too quickly.

But more than that, they would rehearse you with all kinds of questions and prepare you to look for traps set by the OC. If a session with you on the stand does not go over well (for them) they will corner you later and do a post-mortem on every word you said and did not say, and frankly say you disappointed them and came through as weak, even unsure.

Naturally, you have to say you will try harder the next day ... and you certainly will.

Even if you are experienced, each legal team has a different style, and WC will want to tell you all about it – it may be good for you to listen and not say, *"I know, I know ... "*

Most of all, WC is at his best at 'damage control', if and when you mess up their strategy.

As EW, you believe you have built up an unassailable edifice of unshakable science in your client's favour. Alas, this rarely happens.

To put it bluntly, you will be lucky to get off with half your testimony intact. After all, the other side's business is to demolish your testimony, like the boxer's goal of downing his opponent in the first round.

By his crafty questioning, OC may bring out a few inadequacies and even some inaccuracies in your investigation. You may say some things which could be ambiguous or vague.

You grow more and more frustrated and nervous, unable to explain yourself at length like you can in class or in a meeting, but simply having to dodge the darts thrown at you by OC, until the blessed moment that OC grinds to a halt, having exhausted his quiver of arrows, which too, naturally gets replenished each day by his EWs and his client.

The only hope for salvation lies in WC's second round with you, the re-direct or re-examination This is purely to dig you out of the pit you dug for yourself ('accidental' though it might be) so that the fallout of your folly may not destroy your client's case completely.

Once OC has finished your cross, WC will try to get an adjournment, or at least ask for a break to confer with you. Then he will demand how come you did not know what the gaps, loopholes and weak-points were. You have to apologise and promise not to do it again!

WC must now patch up holes blasted by the other side in his ship to save it from sinking. A list of questions must be drawn up through which you may clarify, expand upon, or even correct the damaging statements you made during the cross examination.

During re-examination, your lawyer may read out what you said the previous day, and ask *"What exactly did you mean by that?"* to give you a proper start. This works well for the case. But your lawyers may not go out of their way to correct all the wrongs done to you, unless it helps the case. Court time is too valuable to indulge in ego trips!

4.5.4   Expert Witness's support to Witness Counsel

In turn, EWs have to help lawyers build their case. Lawyers are not doctors, engineers, or accountants, unless they have specialised in one of these fields. They must be briefed on the technicalities, and naturally they look to you for guidance and briefing on terminology, principles, and critical details which will come up during testimony and examinations.

Remembering you cannot introduce any new piece of information or visual aid or model on the spot, you must make each statement self-explanatory and self-contained.

You must brief your counsel in simple and clear terms, with sketches and even models, as necessary, not only for your work but also for the opposing side's testimony.

They should be given a list of questions with the right terminology to ask you in direct examination, to elicit what you want to emphasise, and another list to debunk or embarrass the other side's EW.

You may be asked to assist them in Court during others' testimony, which you do sitting behind your lawyers' chairs and passing them slips of paper with your notes.

That can become hit or miss. If your lawyer senses your comment is critical, he can take Court's permission to confer, then huddle and talk in low voice, or take a five-minute break, go out and talk. When lawyers take a break, the judge retires to his chambers, and sometimes gets involved in some phone call or deep study, and the five minutes may extend to more.

More often, the paper slips are received by the junior sitting next to the questioning lawyer who is on his feet and has only a few seconds to glance at your slip while the witness is answering his question.

If during your earlier discussions going over potential scenarios you had happened to include the point on which you have written a note, then the lawyer catches on easily and uses your suggestion. If he does not understand and cannot clear his doubt with a whispered question to you, he just has to give your suggestion the go-by.

It is indeed a weighty assignment which, if practised seriously and the briefing is thorough, can influence the direction of a case. So the EW has to be on his toes not just about his part in the case, but also about others' testimonies, as and when they are going on.

Personal Experience – 4:

In one case, I had to face off against an EW (say Mr. X) with high credentials. I cannot talk about the exact episode, but the gist of it was that I got a lot of criticism from OC through Mr. X regarding the approximations of a certain dimension which I had estimated by computer graphics similar to what I have described in another paper, [4.3].

But when Mr. X's turn came to be on the stand, my lawyer, who had been briefed by me, was able to turn the tables on him by pointing out that his estimate of a similar item in his testimony was just his guess with no benefit of actual measurement or graphical estimate.

(4-End)

4.5.5 Expert Witness's support to the Court

The EW, being a non-litigant, has a duty to support the Court when the judge requires extra information on the material presented by him, or even on something related to it.

Personal Experience–5:

During my EW testimony in a case involving rebar grid support failure the judge wanted me on two occasions to come up with information on the effects of the loss of one or more such supports, and of staggering of support alignment. I was able to come up with the required results and present them the next day.

I have detailed this episode in another of my papers, [4.3].

Each time, the judge thanked me for the additional information.

(5-End)

But these extra duties were not a favour to the Court. The judge had a duty and right to get more information from EW – who might not have done a good job in the first place. In fact I am not even sure that all the new information helped our case.

My chief satisfaction was more that as an expert, I was able to produce what the judge asked, without having to ask for extra time. Otherwise my credibility as EW would have been damaged.

4.6 EXPERT WITNESS TESTIMONY

Almost universally, the Court is a place of strict decorum and rigid protocol. Everybody addresses everybody else in very polite, normal tones, regardless of the seriousness of the content or the heat of the argument.

In U.S. courts there is often some theatrics, but in most other places, there is very little drama, except occasionally when the witness is actually or apparently sick or overcome by emotion, especially anger or sorrow.

4.6.1 Preparation

The first time an EW takes the stand can be quite unsettling, even for professors who are so used to facing big audiences. You are not accused of any wrongdoing, but you will still face severe questioning regarding your entire professional life and your accumulated know-how for which you are so admired by your colleagues, relatives, friends, and well-wishers.

If you do well on the stand you will come out smelling roses. But if you do badly, you would lose a lot of credibility. Your lawyers help; but inside the Court, you are on your own.

Review the entire case and your part in it. Try to rest well the night before the testimony.

4.6.2 Taking the stand

Come well dressed, but comfortably – don't fashion up in the latest shirt if the collar is a little too tight. You will have to loosen it during your testimony, and it will look sloppy and to critical eyes become a display of your nervousness.

Apart from a lunch break, you get a coffee break once during the morning session and once during the afternoon session, and any time the judge has need for a break. As EW you can always ask for a glass of water, a toilet break, or a recess if you are tired or hungry.

Figure 4.1 is my version of a typical courtroom – photographs not being permitted in courts – but countries and different level courts may have their own arrangements.

In the picture, EW is at left in his cubicle. Court reporter and clerk sit in front of judge's table. The two legal teams have their separate spaces. A cart for the many files, and projection facilities for approved visuals are shown. Portraits of state head and spouse, and a flag or two may adorn the judge's podium.

Fig. 4.1 Typical courtroom for expert witness

You will be led to your seat, and all documents will be ready at your table. After both legal teams have taken their places, the judge's entry will be announced (– '*All rise!*') and all will stand. The judge bows to the assembly, the assembly bows in response, and all sit.

Once the judge signals for proceedings to start, the clerk announces the case, and you will be asked to recite the oath to "*Speak the truth, the whole truth, and nothing but the truth*".

You are not taking an examination. It will not be a memory test – quite the contrary, you will have a surfeit of information. You will have all statements made by everybody and all documents connected with the case in a number of fat files on your table, and the lawyers will be very helpful in your locating a particular page. If you like, you will be permitted to make brief notes while you are being questioned, to enable you to answer coherently. Beyond this, if you wish to use any fresh material from the stand, it will have to be cleared in advance.

Personal Experience – 6:

In a certain case, after I had distributed the handout including copies of my slides for my next-day's testimony, overnight I inserted one fresh last slide in the belief that it would present a summary of some previous slides and elucidate the point I was trying to get across.

Next morning when I showed it on the screen, there was a rustle of papers from everybody searching for its printout, and the judge looked up at me and said, *"Professor, it is not in my set.,"*

When he found nobody else had either, he said to me, *"You are not supposed to show anything without prior approval, please take it off the screen."* Then he told the rest, *"Please disregard this slide,"* and looked back at me and said *"You may proceed."*

I felt a little foolish, but it was not a body blow. I skipped the slide and apologised. The mischievous part of me noted: 'But everybody got to see it, they cannot un-see it!' Of course, I cannot play that trick consciously anymore, because the system keeps track of such things.

(6-End)

4.6.3   Acceptance of credentials

Your acceptability as EW will have to be established in Court before you will be allowed to present your testimony. Your WC will be quite gentle and straight with you. But when OC gets his chance, he will go all out to disprove your credibility – that is his right, even duty!

Theoretically, the OC may direct questions on your competence only on the five key attributes of an EW listed earlier: Education, training, experience, skill, and knowledge. But OC will manage to squeeze in extraneous and personal matters through crafty questioning.

When you submitted your detailed bio-data and supporting documents, you should not have selectively left out sensitive items. If you had flunked a course, or if you had been fined for a traffic violation, you should have put it in.

The other side is going to find out anyway, especially in these days of computerised records. It may not affect your expertise for the case (unless it happens to involve your past indiscretion) but leaving it out would raise questions!

ASCE, [4.2] lists the common criticisms of an EW's qualifications by OC as follows:

- Experience not strong enough in the particular field of concern;
- Training not adequate in the particular field of concern;
- Training does not encompass all the subjects to be covered;
- Too young to have the experience needed to qualify as an expert;
- Has connection with the case and therefore may have a bias;
- Has overstated his qualifications;
- Is a 'professional witness' [or 'hired gun'] whose opinion may be suspect;
- Works only with one type of client;
- Works only for one side (plaintiff or defendant) all the time;
- Has had an adverse conduct reported to licensing authorities;
- Too academic, has not designed many actual constructed facilities;
- Publications or talks not objective enough for the present case;
- Publications or talks contradict his opinions in the present case;
- Does not have sufficient knowledge of applicable standards;
- Not knowledgeable on recent developments; and/or,
- Has failed to stay informed of changes in practice.

It would be the duty of WC and judge to monitor the fairness and applicability of OC's questions. Since no expert can satisfy all the criteria implied, the judge will have to permit non-compliance or partial compliance in certain matters as sufficient for the purpose stated.

Naturally, in case too many of the above charges stick, then the EW can be disqualified, or even impeached on a charge of conflict of interest or falsification of credentials.

More often than documented facts, claims of your experience in relation to the case come up for some juicy questions.

Personal Experience – 7:

I cannot give details, but let us say I am EW for fall of a tourist. OC asks, *"Professor, before this, how many falls have you investigated of a 60-year old 6 feet tall 200 lb. American lady, from first floor balcony of Hotel Excellent?"* [I exaggerate somewhat, but not much!]

I start *"None, but ..."*, and I am immediately cut short by, *"Please say 'yes' or 'no'!"*

If I were given a chance to proceed, I might explain that my experience covered detailed study of numerous falls in different countries, both genders, widely ranging in age, height and weight, falling from different heights and from various buildings and locations.

But the exact combination of the six variables specified? What are the odds, billions to one?

I thought I would say *"My expertise is on falls from height ..."* and attempt a wise-crack that I was not an expert on over-weight American (or <u>any</u>) ladies – but wise-cracks are a 'no-no' from EWs. Who will believe that an engineer can be witty in addition to being expert?

The judge asks the lawyer to rephrase the question. Everybody understands that it was just a rhetorical question to de-stabilise and upset me so that my subsequent responses will be resentful – but I didn't fall for it. To the rephrased query of how I felt competent to testify on this particular case, I responded with what I planned to say, leaving out the joke, of course.

(7-End)

Personal Experience – 8:

Almost every time, OC first starts charging me with not having relevant qualifications to be an EW in that particular case, and I have to pick myself up and climb patiently step by step to the plateau which would be my proper stage for delivering my opinions on the case.

- ■ *"You have spent 40 years in structural engineering, and only 15 years in workplace safety. How can you be an expert in accident investigation?"*
  - I cite my consultancy experience in the USA relating to accident investigations, and my publications on the topic which confirm my contributions to safety.
- ■ *"All your workplace safety experience is in construction. How can you be an expert in* [say] *airports?"*
  - I point out that my expertise and hence my testimony will cover only the topics common to construction and airports, such as scaffold safety.

(8-End)

4.6.4   Technical twists

Quite frequently, the questioning becomes highly technical. Apart from explaining complex scientific principles in simple terms, you must make sense to the people not in your field.

Personal Experience – 9:

In one case, after I had explained my computer structural analysis, the other side argued that I had not considered the self-weight or dead load of the structure and hence my analysis was not valid. I was able to explain that my analysis was to point out the other side's error only in the matter of live loads and its influence on the results.

They then claimed it was impractical and unrealistic to omit the self-weight from the analysis. I explained that most computer matrix and finite element analysis packages actually solved for dead, live and other loads separately, and then internally took linear combinations of the separate results to get various combined solutions for the actual loadings.

Surely their structure/computer EW knew this? He had been my student after all!

But, it was their duty to ask – and it was mine to answer, without getting flustered. After all, I knew. Their EW knew I knew. I knew he knew I knew. No harm done.

(9-End)

So, remain within your bounds of expertise. Being tripped up in Court is not pleasant!

OC will find a weak spot, or some past episode to embarrass you with vague innuendo. In particular, he may bring up some setback you may have had in your previous case(s).

It sounds unfair, but as long as you have stuck your neck out, you must take it on the chin! Simply concede the facts, request permission to clarify with more details, or to explain extenuating circumstances that landed you in that situation. But it will all sound a little lame.

I don't blame OC. It is all in the game. My lawyer will do it too to the other side's EW!

4.6.5   General guidelines for testimony

Some additional guidelines for EW testimony may be listed as below:

- Do not provide interim reports.
- Do not agree to be an advocate for one side or the other.
- Avoid oversimplification and misleading exaggeration.
- Separate opinions from facts.
- Concede indisputable facts even if adverse to client.
- Do not accept assignment without examining relevant documents and reviewing background information.
- Avoid assumptions wherever possible. Try to get information on them.
- Consult other experts or their publications.
- Terminate agreement if client will not let you proceed in the direction you want.
- Express opinion when founded upon adequate knowledge and

honest convictions.

- Testify on standards of care only with reference to what prevailed at the time in question, and also not to suit personal preferences.
- Avoid predicting the future.
- Be dispassionate when rendering testimony or during cross-examination.
- Comment on another's work technically, but do not criticise.
- Report unethical or illegal behaviour of any individual or organisation to authorities.

4.7 ANSWERING QUESTIONS

4.7.1 General guidelines for answering questions

Mostly you will be questioned by lawyers; occasionally the judge may ask a question. You should look at the questioner calmly and impassively, not frowning or screwing up your face in concentration or consternation. (Practice your facial expressions in front of a mirror!)

When a lawyer asks you a question, you do not address him directly; look at the judge, address him as *"Your Honour"* and deliver your answer or comment. If you need repetition of clarification of the question, you may ask OC or WC, addressing him as 'Sir'.

If you are the prosecution or plaintiff's EW, the WC starts questioning first, in the 'Examination-in-Chief', where he takes you through your findings and comments on behalf of your client. If you are EW for the defence, your turn comes second.

When you take the stand, do observe the following guidelines – there are a lot of them:

- If you do not hear a question clearly, ask for it to be repeated; if you do not understand the question, ask for it to be rephrased.
- When the question is in many parts, make brief notes of the different parts, and answer them in sequence.

- When question is ambiguous or (from OC) obviously tricky, ask for a clarification.

- Do not answer an unclear question, interpreting it according to your own understanding, because that may not be the original meaning and it may reveal some information to the other side which they did not ask for.

- Believe firmly in your own testimony based on your knowledge and your investigation and findings.

- When the question probes beyond your current knowledge or competency, admit not knowing the answer, or not being able to answer without further investigation, rather than guess from the stand. Say simply, *"I don't know."* Don't hedge beyond that.

- Keep answers as simple as the context would allow.

- Use drawings or photographs, slides or videos to illustrate your answers – these days courts have excellent opaque, transparency and slide projection facilities.

- Avoid vague or highly jargon-laden answers.

- Answer any question put by OC, unless WC objects and the judge sustains the objection, or the judge himself stops the current line of questioning.

- When OC is aggressive or rude in his questioning, do not get upset or retort or argue. Relax, remain cool – count up to ten – keep looking at the Court with a calm face (no bathos, no martyr look!), and answer briefly and politely.

- If OC persists in straying too far out of line in the questioning too many times, the judge may ask for a meeting of both lawyers in his chambers or at the 'bench' (his table), where he may warn one or both of them to stay within limits.

- When disagreements surface between you and OC, do not start debate or argue, but back-off with the statement "I do not agree", retaining cordiality and decorum.

- Never say *"I feel ..."* or *"I guess ..."* – your feelings and guesses are not of interest to the Court. Say *"My opinion is ...",* or *"I believe ..."*

- Witnesses are not to argue with lawyers; they must restrict themselves to replying, *"No, that is incorrect"*, *"I strongly disagree"* or something to that effect; that is all.

- Do make notes to later tell WC how to counter OC's charges so that WC can ask you the right questions during his re-examination after cross.

- Again, you should not be concerned about your client winning or losing, but only whether what you said in Court is upheld, left alone, or shot down. If your explanation is shown to be wrong, counter it if you can, learn from it when you cannot.

Remember, you are not the accused; what have you to be afraid of? But you chose to get into the ring with the other guy – you can't cry about getting punched once in a while!

ASCE, [4.2] gives the following specific additional suggestions:

- If questioning has strayed beyond point at issue try to bring back the discussion to the relevant point, focussing on the technicalities and not on the personalities.

- Emphasise the professional unbiased way you went about your analysis, not by reciting your virtues, but listing the procedures to take care of possible bias.

- Make your response in obviously lower key than the question, so the OC will be shown up to be the escalating element and not you.

- Avoid terminating the communication; go along as long as you need to for the OC to run out of steam, or the judge to stop the slaughter.

- Don't be cowed into admitting the OC's contention, unless it is true. Be flexible, concede that other viewpoints are possible, but (if you are still convinced of your rightness) reaffirm your opinion, gently but positively.

4.7.2   Take it slow and easy

ASCE, [4.2] suggests that during questioning, you take your time answering. Listen to the question patiently with attention so you don't miss anything. Ponder over the question and consider possible answers. Frame your choice the best way you can, and deliver it calmly.My lawyers have advised me, and warned me, many times. I am learning!

Personal Experience – 10:

I found I was frequently answering too fast. Even during normal WC questioning, I would start to answer before he had completed his question. He would bear with me as best as he could, but tell me during the first break that I should wait until he had finished the question.

But when I cut off the OC's question in a similar fashion, he would look at the judge, turn to me, hold up his hand (which would stop me) and say: *"But I have not finished my question!"* I had to stop in mid-sentence which made me look foolish. My lawyer got quite upset about it and told me it might affect the outcome if I didn't watch out.

I analysed myself and found that this was a teacher disease: 'I do know the answer, so let me deliver it quickly, without wasting time!' That problem threatened to become a bad habit.

So these days, I wait until the lawyer has completed his question. I let him wait until I count up to, well, say up to five, and then look at the judge and speak my response in measured tones different from the rushed torrent of words I used to deliver earlier.

(10-End)

4.7.3   Answer to the point

Answer only to the question asked, and not volunteer information beyond the question as phrased – if the lawyers or the judge want more information, they will ask for it.

Personal Experience – 11:

In a case in which I was an EW for the prosecution, a reputed engineer had just finished his testimony for the defence, explaining how he had checked the scaffold design calculations as per the owner's requirements, and the judge discharged him from the stand. As he was stepping down from the stand, he smiled at the gathering and said, quite audibly, *"I knew that without bracing the scaffold would fail."*

Prosecution counsel immediately addressed the judge saying, *"Your Honour, may I recall the witness to the stand?"* The judge so ordered, and the engineer and took the oath again.

Counsel asked him to repeat the statement that the engineer had made while getting off the stand, and not realising the trap he was falling into, the engineer repeated it.

Counsel: *"If you knew the scaffold would fail without bracing, why didn't you tell them?"*

Engineer: *"Bracing was outside my assignment to check. Moreover, nobody was paying me to ensure the strength of the erected scaffold!"*

It would take too long to describe the aftermath of this bombshell! Suffice it to say that under prevailing rules at the time he could not be connected to the failure.

(Now things could be different.) But his not pointing out the potential weakness of the structure fell short of the 'duty of care' he owed his profession – it might have saved the two lives lost in that accident.

He was lucky; he just lost some face, and maybe some business. Under certain systems, it might have triggered a spate of compensation claims naming the engineer as party to failure!

I too tended to over-answer when I started – also a professor's disease, answering one-line questions with ten-line answers. That day I also learnt over-answering can be dangerous!

(11-End)

You usually answer 'yes' or 'no'; 'that is correct' or 'that is not correct'; 'I agree' or 'I strongly deny it', plus a short sentence before counsel cuts you off. You may continue if:

- WC likes the way you started and gives you a chance to place your comment on record until OC cuts you off; or

- OC likes the way you started and wants you to dig your own grave, which you nicely do until your lawyer sees the trap and manages to stop you by raising an objection; or,
- Judge likes the way you started and he thinks it will throw more light on the case, and he silences both counsel and asks you to complete whatever you are saying.

Do you see a trap in this? Or do you see it as your chance to enlighten the audience?

It is your choice, and the outcome may decide your fate.

If you get a chance to explain a complex item, keep it short and sweet – let them ask for more if they want.

Remember, the judge can stop you cold any time he wants. You are not the keynote speaker at a conference or a minister at an opening ceremony!

4.7.4 Keep your cool

EW is not supposed to get emotional. He is expected to be cool and dispassionate, not a nervous wreck on the one hand or a pompous and arrogant loudmouth on the other. Some can handle it naturally; some should never be an EW. But many can deliver their stuff with aplomb and dignity with some tutoring and determination (and of course with practice).

That is why the OC will push you to the limit, often until WC and/or judge stop him.

It is important for EW to expect to be pushed thus, and be ready for it. Don't scowl in anger at the OC; don't hang your head down in frustration.

At the other extreme, don't smile or grin as if you are a champ winning a round. Answer simply, briefly, politely, firmly.

The usual trick for the OC is to impute atrocious motives to the EW on the stand, such as *"I suggest that you framed your testimony according to your client's instructions."*

Of course he expects you to get shocked and blurt out some hasty rejoinder, which, while it may not be un-parliamentary or profane, will still show you in a bad light.

Personal Experience – 12:

In one case, the smiling OC said to me: *"Sir, I put it to you, that you went into this case with the pre-conceived notion that the component was wrongly designed and that is why it failed, and you planned all your analyses and built all your arguments to support that a-priori conclusion, didn't you?"*

With judge's permission, I explained how I explored many other alternatives by various scenarios, and only after examining all findings had I concluded that under-design was the reason and not over-loading or poor construction. Actually OC's curve ball had helped me!

(12-End)

4.8 COURT ETIQUETTE AND DECORUM

In movies and TV serials, heroes and villains alike spill out their emotions and give vent to their feelings while on the stand, making speeches and acting out. But that is not real life!

4.8.1 Polite and formal

In the U.S., while most testimony is quite polite and disciplined, there are occasional fireworks, by the accused who hurls invectives against the lawyer, or plaintiff who jumps into the Court 'well' (the space in front of judge's podium), or by a lawyer badgering a witness until he breaks down, or by a client or visitor from the gallery shouting and throwing things.

But in actual fact and in most countries, including Singapore, courts are quite staid, dignified and decorous, even when the case involves high-level fraud, or low-level sex.

People are usually dressed formally, although blue-collar workers and young rebels come dressed as they please – latter warned by their lawyers not to look too outrageous on purpose. If the witness does not speak the local language, or is not sufficiently educated to understand and respond in the proper terms, an interpreter is provided to translate back and forth.

The judge has absolute control over the Court proceedings. He can expel anybody at any time if he does not like their attitude or behaviour. The ordinary witness may be allowed some latitude in recognition of his nervousness and lack of familiarity and training about Court behaviour; but EWs are expected to be well-versed in courtroom etiquette.

I had spent a long time in USA, where mild laughter at an unintentionally funny response or a titter at a silly error was permissible and often heard. But in British courts, and in Singapore courts modelled after their British counterparts, frivolity is frowned upon.

Personal Experience – 13:

Once, when I was SW in Singapore Court, sitting just behind my lawyer to help him, I was relaxed, too relaxed because my testimony was over.

The witness on the stand said something really absurd. In the pin-drop silence that followed before the lawyer rephrased his question, a titter sounded loud and clear, and it came from my throat. Everyone glared at me.

When we came out for a break, my lawyer took me to a corner, fixed me with baleful stare, and said in a low but unmistakable growl, *"Look Professor, one more sound like that, and one of us will be out of the case!"*

I too looked him in the eyes and said, *"But I was only clearing my sore throat!"* I think he believed me, but he had the last word: *"Next time, please go out and clear your throat!"*

(13-End)

4.8.2   Sign language

To circumvent the rule against coaching the witness during testimony, I have known of (American) lawyers, clients and witnesses developing a sign language of their own beforehand, such as a lawyer leaning back and scratching his nose to mean *"Say 'I cannot recall*''''. I am sure many others too would like to do similar things.

Knitting of the brows while concentrating and other natural reactions, and occasional involuntary ticks may be overlooked, but most lawyers and judges know all the other tricks that a witness can adopt on his own or after some tutoring.

So, when you are in Court as EW or part of legal team, don't scratch your head, tug your ear, rest your head on your cupped hands on table, or pensively tap your nose with your pen.

What is also not allowed is for visitors in the media or public areas inside the courtroom to react audibly or visibly to anything going on between the principal actors in the Court.

4.8.3   Watch your words

During any break, EW is not to communicate with his lawyers about the case. Generally the judge will admonish the witness against talking to others on the case.

If you are within talking distance from your team in the corridor or at the urinal, you may trade pleasantries of course. Remember someone from the other side may also be there. But lawyers may (and do) kid each other any chance they get – sorry, but you are not part of it!

Everything you say and do inside a courtroom is automatically on record.

The only way you can alter any of them later would be to seek the judge's permission and submit a corrected version. But if it involves alteration of previous evidence, it may become a bigger issue, and consequences may range from an admonition to dismissal of the case.

In the beginning, I wondered about how anybody will know if two people exchange ideas secretly.

Here enters the concept of professional ethics which in effect says that the strength of a rule depends on everybody following it faithfully – hence, for sheer survival of the system, everybody follows the rule. This principle applies to all professionals, of course.

Further, any time someone is found to have broken it, the retribution is swift and extreme.

You must be extremely careful to speak to team members on your side only when asked, and even then not talk loudly or freely – what you say in the Court premises, or even outside in the presence of witnesses, may be cited in the courtroom by the OC.

4.9 ETHICS IN FORENSIC ENGINEERING

Ethics in forensic engineering is the same as ethics in any other field of engineering, but there are a few issues which are unique to forensic engineering:

- Contingency fee arrangements whereby the engineer stands to receive a higher fee if his client wins is unethical.
- Ethics boards are specially cautioned in investigations of complaints against forensic engineers to eliminate harassment, vendetta, etc. by the losing parties.

ASCE, [4.2] documents unethical practices by forensic engineers and EW to include:

- Exaggerating the influence of one factor in a set to explain some failure without considering alternative explanations, or the contribution of other factors;
- Agreeing to limit the investigation to a very narrow scope which the client and/or lawyer believes will help their case;
- Removal of evidence from, or altering evidence at, the scene;
- Deleting or editing non-supportive information from documents, exhibits, etc.;
- Deleting portions of data from records;
- Showing only selective supportive data in graphics;
- Altering or discarding photographs or videotapes;
- Being purposely vague or ambiguous about ongoing work or in stating opinions; and,
- Withholding relevant material during discovery, such as:

o Excluding pertinent background or previous experience unless asked;

o Failure to declare who did the actual work;

o Wilfully ignoring factual data;

o Attempting to recant or correct previous testimony when subsequent developments do not fit them; and,

o Manipulating the resume to make it look better, or to mislead the reader.

Borderline unethical practices include the following:

▪ Volunteering answers or information beyond the scope of the question;

▪ Exaggerations, sarcasm, jokes, guffawing;

▪ Pontificating, giving unnecessarily lengthy answers;

▪ Criticising another engineer's work;

▪ Inspecting other's exhibits without their permission;

▪ Discussing testimony during recess with persons other than client or client's counsel;

▪ Arguing or verbally warring with lawyer or judge; and,

▪ Openly displaying approval or disapproval during other witnesses' testimony.

The lesson to be drawn from such a situation is that an EW must weigh the ethics of all aspects of a case and feel comfortable before accepting an assignment or acting on it. If it is his misfortune to joust against a less ethical EW, he must have the courage to stand up to the latter, and/or make a complaint to the relevant ethics committee after the case.

Lawyers' ethics also forbids expecting an EW to violate his professional ethics.

Personal Experience – 14:

While I was on the stand, I was once asked: *"What do you think of the expertise of Mr. Y [their EW from abroad]?"* Mr. Y's resume was indeed formidable, but that was not the point.

I simply replied: *"My professional ethics does not permit me to evaluate the credentials of another EW in the same case.*"

(14-End)

That is the way it should be between professionals – each doing his job for his side as well and as ethically as possible, but at the same time respecting other professionals.

Personal Experience – 15:

At the end of both our EW stints after we had torn each other's testimony apart through our respective lawyers, I went up to the other side's EW Mr. X, shook hands and said I was happy to make his acquaintance, which I indeed was because I learnt some things from his deposition while I was highlighting for my lawyers where they should tackle him during their cross-examination.

He was nice enough to praise my website (which of course, he had browsed through during the previous weekend), and we went on to chat about our experiences with the U.S. Government division for which both of us happened to have consulted.

We didn't of course praise each other's performance on the stand!

(15-End)

4.10 CONFLICT OF INTEREST OR BIAS

Conflict of interest for EW can be prior, current, or planned personal or business relationship with one of the parties, an interest in or benefit from the outcome of the case, or anything else that might compromise his impartiality.

The conflict may be:

- *Actual* – certain to affect opinion;
- *Latent* – may have reasonable chance to affect opinion; or
- *Potential* – can be foreseen to cause conflict of interest.

As you do not know all the persons or organisations involved in a case, or which relationship could be a conflict of interest, best thing to do would be to disclose to your lawyers all possible conflicts to the extent you know them, and let the lawyers decide for you. If and when something comes up during the case which concerns you on this matter, you may again confer with your lawyers and let them decide for you.

Personal Experience – 16:

When I was asked to be EW in a case which involved a quasi-government organisation, I explained to the lawyers that I was at that time a research-consultant to another Government department. The lawyers cleared me to go ahead, and I duly informed my research sponsor.

I have also refused cases where I found I might be facing my clients in my consultancy, albeit from another agency, across the aisle in the case.

However, I could not avoid facing opposite me, some of my former students (who by now had risen to quite high positions, about which I could be justifiably proud).

In one case, the EW and one of the defendants on the other side had been my students. Their EW did not have a problem because he and I knew we were just experts and not litigants – although he had not seen me in action as EW yet! Of course he helped his side to punch holes in my arguments!

The defendant however was embarrassed to let me see him in such an awkward position, I guess.

The first day we met in the Court corridor, he greeted me with a shy smile and a *"How are you, Professor"*. As the heat of testimony rose, he simply sat, avoiding my eyes.

It turned out that finally the case against him and his two co-defendants was dismissed.

(16-End)

Bias, according to ASCE, [4.2] is an *"inability or unwillingness to consider alternative approaches or interpretations."* The charge of bias in an investigation can be eliminated or minimized by using more than one method to analyze the data and to test the validity of the possible scenarios. Bias is misplaced sympathy which might sway one's judgement.

The possible consequences of proof of conflict of interest and bias can be many and serious, such as: termination of services by the client, denial by the Court to testify, dismissal of the client's case by the Court, and disciplinary action up to and including suspension or even removal of P.E. license.

4.11 MISTAKES BY EXPERT WITNESS

Mistakes may happen with any EW; this will damage your testimony. Alternatively, while yours is one explanation for the accident, the other side may come up with another equally good or even better explanation which fits the facts; this will weaken your testimony.

The magnitude of the mistake and its consequences to the parties on the two sides may become primary factors in the disposition of the case. Another factor would be whether you find the mistake yourself or the other side finds it. Either way, it is a blow to your side's case.

The legal aspects of mistakes by EW are very complex, and highly dependent on the legal system and many other circumstances. For instance, in certain states, EWs have immunity against mistakes and their consequences. USA abolished blanket EW immunity in 2001.

Charges may cover perjury, contempt of court, malpractice, negligence, etc.

Outcomes may include declaration of mistrial, claims for damages, fines and imprisonment, fresh litigation, etc. Consult a legal expert in forensic engineering for guidance in this matter.

4.11.1 Common mistakes by Expert Witnesses

A general summary of common mistakes sourced from various authors [4.6, 4.7, 4.8, 4.9, 4.10] is given below. Many of them have been identified in earlier sections of this paper also. It is a long list, not in any particular order, but try to avoid them all like poison!

(a) Accepting a case:

- Preparing different CVs for different clients.
- Accepting rush cases that do not permit the engineer to follow his standard protocol.
- Accepting low-budget cases.

(b) Preparing for investigation:

- Failing to adequately document their investigation and findings.
- Failing to ask for and review the complete set of records.
- Not corroborating facts provided by counsel.
- Simply reviewing attorney's deposition summaries instead of reading the deposition.
- Relying on data or documents not pertinent to the date on which the event occurred which has given rise to the lawsuit.
- Refusing to acknowledge the implication of possible bias resulting from prior employment (i.e., pensions, friendships with career co-workers, post-employment contracts, clients derived from prior employment)

(c) Relationship with attorneys:

- Neglecting to tell your attorney about related prior testimony, affidavits, speeches or publications, job assignments or lawsuits even if remotely relevant.
- Failing to reveal "resume blemishes" to your attorney (failures in school, convictions, drug or alcohol problems, job 'lay-offs', conflicts of interest, prior accidents, license suspensions, etc.).
- Not providing your attorney cross-examination questions for use on opposing expert.

- Not ensuring counsel understands the investigation and the findings thoroughly.
- Ignoring the opposing lawyers' and experts' view of the case.
- Misunderstanding how your opinion fits into client's or attorney's theory of the case.

(d) Conducting the investigation:

- Forgetting that you are an advocate only for your own opinions, and your methodology, but not for opinions happening to favour your client.
- Failing to master the facts of the particular case in which you are employed
- Mishandling custody of tangible evidence.

(e) Writing the report:

- Writing reports that are based on incomplete investigations and insufficient data.
- Writing a report without being asked by counsel.
- Putting too much in writing, too soon, and too casually.
- Not writing a report according to civil procedure rules, if any.
- Sharing draft reports with counsel.

(f) Behaviour on the stand:

- Sounding too much like an "expert", that is, talking down to the Court.
- Offering opinions outside your area of expertise.
- Allowing your "ego" to intrude in your deposition or trial testimony.
- Damaging your credibility by quibbling over peripheral issues when on the stand.

- Losing your temper on the stand – except for restrained righteous indignation.
- Answering hypothetical questions without asking the cross-examiner to supply all the variables or assumptions.
- Answering questions in deposition or trial, without understanding the question.
- Losing sight of the fact that juries and judges pay a lot of attention to choice of words.

(g) Fee discussion and billing:
- Billing for work not authorized by the attorney and client or analysis which satisfies intellectual curiosity but is not necessary for the case.
- Revealing arrogance when discussing how much you are being paid to testify.

Professional consequences and management of technical mistakes are another matter.

4.11.2 Who discovers the mistake matters

If the other side finds a major mistake in your work, it would be disastrous to your side.

If you find the mistake yourself, first inform your lawyers. Analyse the error, do whatever you can to correct it, and determine what other findings, conclusions, and recommendations the error and correction would impact.

As soon as possible submit a report detailing the mistake, correction, and its consequences, formally to the Court through your lawyer.

If the mistake affects only your side, your client and lawyer may decide to soften its impact. You may accommodate them to a certain extent, but do not let it become a cover-up.

First off, a cover-up would be cheating the system, defrauding society. Professionally, it would be highly unethical. Finally, it would be extremely illegal, punishable under law.

Discovery of the error may not be that far-fetched. The other side expert already has (or can subpoena) EW's data, and can re-do the analysis; then the error will certainly show up.

The mistake itself is bad enough. Cover-up would make it much, much worse. EW's career would be over. As a respectable member of society, he would be finished for life. His family and his friends would have to face the stigma forever. It is certainly not worth it!

Naturally, it would be very embarrassing to have to confess to a mistake. The plus point here is that you caught it yourself and took the initiative to rectify the error. Of course you will lose some credibility, but people will understand and respect you for it.

Actually, being gracious in defeat would score some points for you in the other testimony where you are right. You will survive, to do better and fight again.

4.11.3   Management of mistakes

Whether it is an outright mistake or dilution of opinion, you must accept the error or the change. Don't let your loyalty to your client, or guilt at taking his money, affect this matter.

- Blustering, arguing, making excuses, joking about it, blaming other people or faulty equipment (computer is the usual scapegoat) is a cop-out, hallmark of a bad loser.

- If there is any explanation of the error in your favour, do bring it out, but do not 'gild the lily' or try to slide out from under. It will only make you look worse.

- Put on a calm face and with some humility, concede the inevitable truth: You were wrong, or the other side is equally right.

Trivial, inconsequential mistakes in your report or Court presentation that would not have any impact on the deliberations in Court or on technical or legal outcomes, may lead only to embarrassment, and correction of record. Revised documentation and apology would suffice.

Personal Experience – 17:

In a recent case, after the preliminaries, when the OC went over my many graphical analyses picture by picture, he pointed two of them which were identical. Sure enough, there they were, for all to see in their copies, and no way could I deny it.

Luckily, I could identify the problem right away: I had two items in one picture, and I had intended to use the same picture twice, once for each item, but had ended up marking both items on the same picture and printing it twice. I thought I had checked everything many times before I handed my report to the lawyers, but obviously I had missed this one item.

Somewhat embarrassing, but once I explained and apologised for my oversight, everybody understood, and the point was dropped. Still, I reprinted that page in the report overnight with the two pictures correctly marked and first thing next morning my lawyer handed copies to all parties, while I delivered an apology with a smile.

(17-End)

If the mistake is in EW's investigation or in findings, the problem is bigger. Even here, if consequences are localised and limited to broad conclusions, and does not extend to primary conclusions and framing of charges addressed by the EW, the damage can be contained.

Personal Experience – 18:

Once, I discovered that I had misread the number from the voluminous output, and had declared the other side's design as *"deplorably inadequate"*. The other side objected to my use of 'deplorable', and asked for more quantitative details for my qualitative assessment.

I had no way of reacting on the spot. Fortunately, it happened about half an hour before end of day, and I requested adjournment through my lawyer.

It was while I was checking documents on why I said 'woeful' that I discovered my error.

Thought of avoidance did not even strike me.

I drafted a succinct and honest reply explaining the circumstances of my oversight, and (after checking as time and scope allowed) setting down how the correction affected my revised opinion (– up to 'barely adequate' from 'deplorably'), and affirming that it did not affect my other findings or recommendations.

Next morning, copies were distributed to all the parties. With the Court's permission, I summarised in simple, unvarnished language what was in the revised document, apologised for this unintentional error and sat down. Of course, I was lucky it did not damage the case!

(18-End)

4.11.4   Consequences of mistakes

Bigger mistakes will have disproportionately bad consequences for the side with mistake.

If the mistake is so bad that it blows your client's case to the skies, or gives ammunition to the other side, that will be really disastrous.

Maybe your side can appeal with a corrected analysis – and possibly look for a smarter EW! You will lose some or a lot of future business.

Once one mistake surfaces, other side will naturally suspect there could be others, and would want re-confirmation from your side, and also have their experts verify your claims.

Apart from the legal consequences (which cannot be detailed here), professional fallout of a serious mistake on the EW would be quite bad. EW's P.E. license will be suspended, and his membership in professional societies cancelled. He may not be EW anymore.

The consequences of mistakes are vastly different from the viewpoint of the engineer and the law.

To engineers, it is clear if in a scaffold collapse case, the designer has been shown to have used the wrong formula or the contractor has been proved to have omitted braces, they deserve to be penalised proportionate to the contribution of their mistake to the collapse.

But the law takes a different view. First the causality has to be proved 'beyond reasonable doubt'.

Most engineers would condemn the use of wrong formulas or omission of braces, but it has to be proved with unassailable evidence that it was the wrong formula or the omission of braces that caused the failure. Next, factors like negligence –meaning the professionals committed the mistakes with knowledge and intent – must be demonstrated,

Further – and this I learnt the hard way – the error must be shown to have violated some Code or regulation in existence, or at least to have been contrary to good standard practice in that particular industry or trade, at the time and in the region where the accident took place.

Otherwise, the most a Court could do would be to warn the proven wrong-doers to be more careful in their professional 'duty of care'.

Defendants or accused would go scot-free. Hopefully, the 'near-death' experience has made them wiser, and society would be better off with that, even without punishment. I have referred to some examples in another paper, [4.3].

4.12 COURT REPORTING

In the old days, the best a Court reporter could do was to take short hand and transcribe it later into regular typed text. Then came, at least in the West and advanced counries, the stenotype machine, which had a few keys with which the stenographer recorded the depositions and testimonies phonetically and then typed them longhand. (Figure 4.2.)

Once electronics entered the game, there were huge changes in the recording modes, with speech recognition and speech-text transscribers, which will do almost everything a stenographer used to do, requiring only a little bit of final syntactical editing, with investment of a fraction of effort and a miniscule amount of time.

But pictures have not advanced in the same way as text has. Photography (still and video) is not yet allowed in most courts, as already noted. Court may record all proceedings in audio and video for its own internal use and archiving, under strict protection of confidentiality.

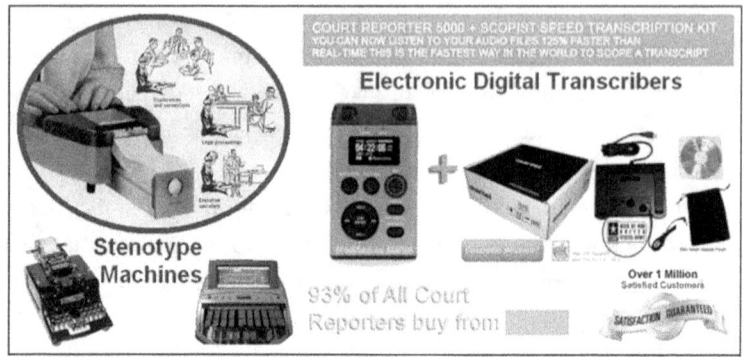

Fig. 4.2 How Court reporters record the proceedings

4.13 DEALING WITH THE MEDIA

All Court investigations of accidents are very public affairs, with rare exceptions like when rape victims, particularly minors, are examined *in camera* meaning isolated from the public.

Often, it is not EW who is focus of publicity but the high profile case with wide public exposure, or if victim, plaintiff or defendant is a high official, TV star, or notorious gangster!

The point is that the media must have its huge headlines to flaunt about the case, interesting stills or videos to display from it, and/or sound bites to quote from one of the players in it. Simply put, the case must sell the paper, or draw TV and radio audiences.

Incidentally, the EW will come in for some unavoidable media attention.

Newspapers and radio/video commentators have to dramatise the issue. They do not misquote, but some of your words may be placed out of context, and displayed in a headline, they may raise eyebrows if not some anger! Use of journalese may give the reader a wrong impression not intended by the person attributed to and not admitted by the reporter; for instance, acquisition of information from the Internet may be cited as 'lifted' from it!

Once the media have their say or show, you don't have a chance to clean up your act in the same media, so you just have to accept it as your *karma* (fate), good or bad.

Personal Experience –19:

In an accident which involved a victim from another country, the case attracted wide local attention, and I as EW fell target to media interest if not frenzy. One morning paper covering my first day's testimony, had headlines featuring the phrase *"International Safety Standards? 'No Such Thing!'"* accompanied by my picture. Could I have meant it the way it sounded?

After second day's testimony, just when our legal team was about to exit the Court house doors, my lawyer straightened my tie, and said: *"Prof, the paparazzi* (that is, media reporters) *are out there with video cameras. Smile at them, walk briskly, but don't say anything!"*

I said, *"Aren't you coming?"* My lawyer said, *"Nah, you are the handsome one"*, and added more seriously, *"We are here every day, and they know we won't tell them anything."*

Nor would I, but the reporters might not know that, so, although I was dog tired, I put on a smile and walked straight past them, even waving away a question, and a couple of videographers followed me all the way to the taxi pick-up.

All four language TV channels (English, Mandarin, Malay, and Tamil) showed me smiling and walking for about ten seconds, while the voice-over explained that Professor Krishnamurthy was the EW for the plaintiff, blah-blah-blah. I don't know what the viewers got out of it, but many people recognised me from that (and I guess have forgotten me when the next, more handsome – or more high-profile – person came along!)

(19-End)

The general rule for EW with media is, better keep your mouth shut. It is for the client or his counsel to announce the appintment of the EW, and the scope of the work assigned to him. Even when the client or counsel encourage or approve the EW's talking to the media, the EW must clearly understand how far they want him to go.

Publication of the EW's work in a magazine or technical journal can be considered only after all the appeals have been exhausted and the case has been closed, and even then, if identification details are to be included, only with the permission of the client.

These days, blogs are the worst to stomach.

Because of the ease of access and the anonymity of their origin, blogs are where immature minds can spill out their innermost fears and prejudices (along with some serious comments and a few rare bouquets).

One blog called me a meddling old man who does not know how to use computers wasting a court's time with his fancy fantasies. You just read and ignore them, that is all.

4.14 THE FEE STRUCTURE

How much do you charge? Whatever the market will stand, is one answer.

But professional ethics will impose constraints such as: Not hiking up the rates depending on the distress of the client; not collecting fees from two parties for the same job; not accepting contingency fees by which he gets more if his side wins and less if it loses.

Generally, accident investigation rates are one and a half to twice normal consultancy rates; Court appearances for taking the stand can be two to three times the normal rate; Court attendance to assist the lawyers may be halfway between the previous two.

So, if a consultant engineer charges $100 an hour normally, he can charge $150 an hour for investigation, up to $200 an hour if site visit, lab work, or out-of-town travel is involved. He may charge $250 an hour for taking the stand, and $200 an hour for advising lawyers.

Of course the rate may go up by another 50-100% if you are known as one who delivers the goods consistently and who can stand up to a barrage on the stand with aplomb.

4.15 THE LAWYERS' ATTITUDES

How about the feuding lawyers who are out to get each other's blood through the poor witnesses on the stand? Outside the Court they will be so nice to each other you wonder if they are the same people you saw inside fighting (politely!) like cats and dogs! Although I have not discussed this with lawyers, privately. I guess they may have their loves and hates, admirations and jealousies, but nothing is overt or intense.

I now compare them not to two boxing champs fighting it out in the ring, but to the Williams sisters Serena and Venus, battling it out on the singles tennis court – surely each would be fighting to win, regardless of the love and the admiration they feel for each other! On court, no quarters given or asked. Off court, back to family life, give and take.

Unlike a doctor or an engineer, when a lawyer represents people who have been wronged, another lawyer out of necessity – even if only because the Court assigns him – represents the wrong-doer. In many cases the line between the wronged and the wrong-doer may be blurred.

An accused may admit to his lawyer any legal violation, but the lawyer cannot convey it to the authorities (or anyone else) under the attorney-client privilege. Instead he would have to defend the accused to his utmost in Court, to get him off the hook.

Even when more evidence comes in against the accused, the lawyer still has to try to get the sentence reduced due to 'extenuating circumstances'. I guess it is like actors who become famous for their villain roles – the test of quality being only *"Did he do a good villain act?"*

In the West, especially in USA, it is common for lawyers at some stage in cases where the trends are fairly clear, to confer with clients and decide what lesser charge they could plead guilty to and accept the smaller sentence, rather than take a gamble and end up with a much heavier sentence.

Then the accused's lawyer meets up with the other side and offers a deal. If this other side also has concerns how much longer it will take or whether they can make their charge hold up till the end. They will negotiate hard until some common ground is found and then approach the judge with their compromise.

In practice, all lawyers are very polite, even pleasant, to witnesses and clients from both sides.

But let there be no misunderstanding, each lawyer is paid to be there (even *pro-bono*, i.e. free social service) to quash the other's case, by destroying the testimony of the other's witnesses, EWs being prime game.

Some OCs may make it clear from start that they are on the other side of the fence from you.

Other OCs act aggressive with you while you are on the stand, but once the judge has retired to his chambers, become very friendly, even chatty.

You may have to be careful with your responses though, not relaxing your guard, and not talking about the case matters.

Most OCs however keep to the minimum courtesies, with a slight smile or nod when your eyes meet, but otherwise leave witnesses well enough alone.

Yet, lawyers can be quite nice sometimes – even to the other side!

Personal Experience – 20:

In a case when I took the stand for a cross-examination on 30th October, I was surprised when OC opened his cross by facing the judge and saying with a smile: *"Your Honour, I think it would be proper for us all to wish the Professor a Happy Birthday yesterday!"*

Judge nods, everybody in Court applauds – a rare departure from rigid daily combative format. I don't think they do it for every witness. I get up, take a bow and sit down again. I know that I should get ready to battle the OC soon! Yet, I am incredibly moved.

(20-End)

4.16 CONCLUSION

To quote Ratay again, [4.1]: *"Expert consulting/witnessing is a challenging, demanding, lucrative area of professional engineering practice. Not everyone is 'cut out' for this work, and not everyone wants to operate in the usually adversarial environment."*

Forensic engineering is somewhat akin to a medical speciality like neuro-surgery. It is very stressful, and very satisfying. The pressures of being an EW are indeed high; but the hazards of failure and consequences from failure are often not worth the risk.

Each time I go as EW and the pressure builds up, I tell myself I should not accept another. But after the case is over I look back and feel good about what little I may have accomplished for someone or some group. Then, I am ready for the next case comes along – if I like it!

4.17 REFERENCES

4.1. Ratay, R.T., "The Forensic Expert Consultant/Witness – Some Things To Know", *STRUCTURE Magazine*, September 2007, p. 58-60.

4.2. Lewis G.L. (Ed), *Guidelines for Forensic Engineering Practice*, ASCE, Reston, VA, USA, 2003, 140 p.

4.3. Krishnamurthy, N. "Investigative Methods in Forensic Civil Engineering", *Proceedings of the Conference and Exhibition on Forensic Civil Engineering*, 23-24 August 2013, Bangalore, India.

4.4. Shuirman, G. and J.E. Slosson, (1992). *Forensic Engineering – Environmental Case Histories for Civil Engineers and Geologists.* Academic Press, New York, 296 pp.

4.5. _____, http://www7.aecforensics.com/wp-content/uploads/2011/06/Recommended-Practices-for-Design-Professionals-Engaged-as-Experts-in-the-Resolution-of-Construction-Industry-Disputes.pdf

4.6.   Thompson, D.E., and H.W. Ashcraft, "Chapter 9 – The Expert Consultant and Witness", *Forensic Structural Engineering Handbook*, R.T. Ratay (Editor-in-Chief), McGraw-hill, 2000, 808 p.

4.7.   Jorden, Eric, *Mistakes Forensic Engineers make,* Retrieved July 2013:
http://www.ericjorden.com/blog/2013/06/20/mistakes-forensic-engineers-make/

4.8.   _____, *Top Five Mistakes Expert Witnesses Make,* Beirne, Maynard & Parsons, LLC. Retrieved July 2013:
http://www.bmpllp.com/publications/17-the-top-five-mistakes-expert-witnesses-make

4.9.   _____, *Top 20 Mistakes an Expert Witness Can Make*, iLaw Connect, 2012. Retrieved July 2013:

http://www.ilawconnect.com/blog/top-20-mistakes-an-expert-witness-can-make?goback=%2Egde_4451478_member_171103524

---

† This is a reprint of the author's paper presented at the Forensic Civil Engineering Conference and Exhibition held at Bangalore, India, on 23-24 Aug. 2013, by the Association of Consulting Civil Engineers (India), reprinted with permission.

AUTHOR'S DISCLAIMER

Two points are worth reiterating lest the reader assume that this is a handbook for expert witnesses by a veteran:

1. Almost all the comments I have made here are from my personal experience. Many of the numbers I have mentioned would be for Singapore aspirants. The fee structure I have suggested would be in Singapore dollars for Singapore beginners. I found (again from personal experience) that the rates for experienced veterans and in other countries the fees may be many, many orders of magnitude higher!

2. I am not a lawyer. I have sat in court in three countries, supported expert witnesses in two, and testified as expert

witness in Singapore. I have shared with readers what I have learnt from my lawyers and other local associates. My description of court procedures and other legal matters may be way off the mark! I make no apology for it, except to confess I may have been too forthright in some of my reactions and perceptions.

So, reader, please take my comments as the casual expression one man's personal experience in walking in a small minefield.

Consult a lawyer if and when you have or want to be an expert witness!

———

5

FORENSIC CIVIL ENGINEERING AND RISK MANAGEMENT[†]

ABSTRACT

Forensic engineering is considered to be a lagging indicator aimed at investigating accidents after they happen, while risk management is usually a leading indicator aimed at predicting mishaps before they occur so as to prevent accidents. In this paper, author attempts to bring together the two apparently opposing concerns and procedures into a single integrated and synergistic whole. Using a number of case studies to illustrate his points, author demonstrates that by a judicial combination of the two specialties, improvements to existing planning procedures may be made to avoid accidents or at least minimize their impact.

5.1 INTRODUCTION

Which came first, the hen or the egg? Do we suffer accidents and regret our ignorance or do we assess risks on the basis of past accidents so we can avoid future accidents? This is a cyclical procedure and which comes first is irrelevant.

What is true today is that we have enough information and knowledge on almost all possible mishaps that can happen to engineering artefacts and their users that we can identify the hazards (i.e. potential dangers) in any activity, and eliminate them or mitigate their effects by risk management (RM).

For good risk management, experience with mishaps and their consequences, that is experience with accidents, is essential. This experience is usually gained from exposure to and analysis of past accidents and their aftermath, in short from forensic analysis of the accidents.

From another point of view, risk management is a leading indicator (meaning that it helps to predict the outcomes of an activity so we can avoid them or reduce their impact) rather than a lagging indicator (meaning that it helps analyse what happened in a mishap after it happened).

Forensic engineering is generally thought of as an accident investigation procedure while RM is an accident prevention device. Although by usage 'forensic engineering' is restricted to accident investigation procedures to satisfy legal requirements, one accepted basic definition [5.1] of the term is as follows:

- Forensic (adj.) *"pertaining to or suitable for courts of law,"* 1650s, with *-ic* stem of Latin *forensis "of a forum, place of assembly,"* related to *forum* "public place". Later used especially in sense of *"pertaining to legal trials."*

- Author would prefer the even broader definition below [5.2]: *"Forensic Engineering is the art and science of professional practice of those qualified to serve as engineering experts in matters before courts of law or in arbitration proceedings."*

These definitions do not preclude the application of the term 'forensic engineering' to pre-accident analyses or in particular to accident prevention exercises. The rest of the paper will deal with the interaction of risk management and forensic engineering with particular reference to civil engineering.

5.2 RISK MANAGEMENT

The basic steps in risk management [5.3] are as follows:
- Identify the hazard in a workplace activity.
- Assess the likelihood of its occurrence and the severity of its consequences.
- Combine their effect by a risk matrix to determine the risk from the hazard.
- The preceding three steps constitute 'Risk assessment' (RA).
- Depending on the magnitude of the risk, check if the existing controls are adequate or what additional controls would be necessary to reduce the adverse consequences of the risk to within acceptable or at least tolerable limits.
- Implement the additional controls.

In general, most accidents occur because the person(s) involved did not identify the hazard, or if they did, did not assess its likelihood and/or severity properly.

Whether all accidents can be avoided by risk management or not, proper risk assessment will direct attention to weak links in an engineering activity and lead to an estimate of the consequences and recommendations on how to control the worst of them.

5.3 LEGAL REQUIREMENTS OF RISK MANAGEMENT

Most advanced countries have stringent requirements for risk management for activities at hazardous workplaces. Singapore has the Workplace Safety and Health Act [5.4] which includes the Risk Management Regulations [5.5] and Code of Practice for Risk Management [5.6].

When the Regulations were promulgated in September 2006,

only the three most hazardous industries in Singapore, namely construction, ship building and repair, and manufacturing were covered. In subsequent years, all the other workplaces were covered so that today, even offices, banks, and barker shops are included in the requirement.

If an employer (or more precisely, an 'occupier' of a workplace) does not carry out risk assessment they will be fined S$10,000 in the first instance, and if repeated, fined S$20,000 and/or jailed for six months.

The 3×3 risk matrix shown in Fig. 5.1 (Top) is the minimum size. The 5×5 numerical risk matrix of Fig. 5.1 (Bottom) is recommended, but any size 3×3 or larger is acceptable.

Likelihood Severity	Remote	Occasional	Frequent
Major	Medium Risk	High Risk	High Risk
Moderate	Low Risk	Medium Risk	High Risk
Minor	Low Risk	Low Risk	Medium Risk

Example of a common 3x3 Risk Matrix with descriptive ratings.

Likelihood Severity	Rare (1)	Remote (2)	Occasional (3)	Frequent (4)	Almost Certain (5)
Catastrophic (5)	5	10	15	20	25
Major (4)	4	8	12	16	20
Moderate (3)	3	6	9	12	15
Minor (2)	2	4	6	8	10
Negligible (1)	1	2	3	4	5

Recommended 5x5 Risk Matrix with numeric ratings or Risk Prioritisation Number.

Fig. 5.1 Top: 3×3 Qualitative Risk Matrix;
Right: 5×5 Numerical Risk Matrix

Definitions of the levels of likelihood and severity are given to enable the assessment of their level by the risk assessment team. From these levels, the category of risk is chosen either from the qualitative matrix as shown in Fig. 5.1 (Top), or by multiplying the likelihood and severity level numbers and determining their risk index as shown in Fig. 5.1(Bottom).

For all the apparent logic and rigour of the assessment procedure, the estimates of likelihood and severity levels are quite subjective. In particular, the likelihood of the mishap occurring is mainly a function of local circumstances and human factors, and its assessment is generally done on the basis of relevant statistics to the particular industry, site, and job.

5.4 INTERACTION OF RISK MANAGEMENT WITH FORENSIC ENGINEERING

As already mentioned risk management, being a leading indicator, is the antithesis of accident investigation which is a lagging indicator. The former aims to prevent accidents and the latter focuses on analyzing accidents. Both must meet legal requirements.

However, we may use both of them synergistically, supporting and reinforcing each other, to accomplish two objectives:

(i) To analyse an accident by risk management principles so as to learn lessons from it – as is conventional practice; and,

(ii) To prevent accidents by conducting a risk assessment before starting a project and implementing required additional risk controls.

The chain of actions may be designed to learn from past accidents to prevent future accidents through risk assessment and control, as follows:

- Determine the immediate and root causes of the accident.
- Check if these causes had been identified earlier in the risk assessment.
  - o If not, ensure that the fresh knowledge is utilised in future risk assessments.
- Check if the actual time frame or frequency of the accident fits within the likelihood assessed for the accident.
  - o If not, update the criteria for deciding the likelihood to include the fresh knowledge.
  - o For example if as in Fig. 5.2, forklift accident frequency of 1 per year from 1999 to 2002 has led to a 'Low' likelihood, when in 2003 there were 6 accidents, it would be prudent to raise the likelihood level to at least 'Medium' as a predictor for subsequent years.
  - o In particular, it may be noted that to reduce the likelihood to 'Low' again in 2005 because there were no accidents in 2004 would be unwise as the three-year average for 2003 and 2004 are still high.

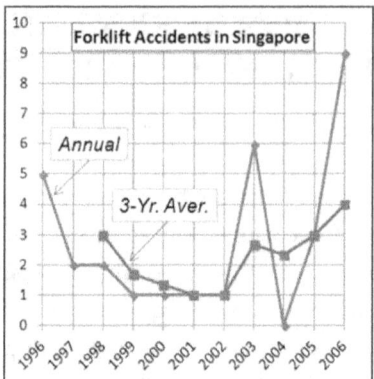

Fig. 5.2 Lessons from Accidents

- Check if the actual impact of the accident fits within the severity assessed for the accident.
  - o If not, update the criteria for deciding the severity to include the fresh knowledge.
  - o If in the example of Fig. 5.2, the worst severity of the increased number of accidents happened to be simple fractures ('Medium') rather than death ('High'), then there would be no need to increase the severity level despite the increased number.
  - o If all of them happened to be just bruises, then it may also be acceptable to reduce the severity from 'High' to 'Low'.

The usefulness of forensic engineering to risk management is well established and widely utilised. It will not be an exaggeration to say that if it were not for forensic engineering, risk assessment will be almost impossible.

It is the investigation of accidents, documentation of details of their severity and frequency, and identification of root causes and contributory factors that build up the database and statistics from which the likelihood and severity of hazards can be assessed with any degree of certainty for future use.

Thus, an industry which has high volume and good quality data on workplace accidents can provide sufficient information to risk assessors to estimate accurately the likelihood and severity of various mishaps that may happen in any workplace activity in that industry.

5.5 IMPORTANCE OF RISK ASSESSMENT TO FORENSIC ENGINEERING

Out of the more than hundred Singapore accident reports in public domain recently studied by the author, about three-quarters had the citation *"Lack of or inadequate risk assessment"* as the first item in the list of possible causes, and/or *"Conduct proper risk assessment"* as the first recommendation to be followed, signifying a weakness in that regard in the occupier company for that accident.

It is also a fact that a cause of most of the accidents can be traced back to the omission of the particular hazard during the risk assessment even when it has been done. So it is clear that risk assessment can be a very powerful tool to anticipate and prevent or mitigate the effect of accidents.

This approach is gaining more and more importance in forensic engineering because in Singapore and many other countries one of the leading questions asked by the prosecution or plaintiff in cases involving workplace accidents is if the risk assessment had included the item found to have been a root or contributory cause for the mishap.

5.6 ROLE OF RISK MANAGEMENT CODES IN FORENSIC ENGINEERING

As outlined earlier, risk management code of practice normally prescribes:

(a) The procedure to be adopted to arrive at an assessment of the risk of a particular hazardous activity; and,

(b) The sequence of consideration and implementation of control measures to be adopted to eliminate or mitigate the consequences of the risk.

These are legal requirements, and violation of any of the mandated terms attracts definite and significant penalties such as suspension or revocation of licenses, fines, and jail terms. Whether a certain party in an accident followed these procedures for risk assessment and control hierarchy will therefore become critical in any court proceedings on the accident.

The author, like most expert witnesses, has faced this question on behalf of his client or has put this question (through the attorney) to a witness from the opposite party.

A court hearing on an accident may include the following line of questioning:

- *Did you conduct a risk assessment (RA) of this activity?*
- *If yes, did your list of activities in the RA cover this particular trigger or contributory event?*
  - o Eg. Reaching beyond the worker's grasp (beyond 2 m) while working at height on a safe base, such as for fixing a light bulb at a 6 m high ceiling from a 3 m high mobile scaffold.
- *If yes, did you have a Safe Work Procedure (SWP) for this activity written out and distributed to the personnel concerned?*
  - o Eg. Provide some means of reaching higher from the work platform, say by means of a 2 m ladder.
- *If yes, did you identify the hazards (potential dangers) in this SWP?*
  - o Eg. Once the worker climbs on the ladder, he will no more be protected by the 1 m high guard-rail specified in the Work at Height code. So further SWP must be specified to prevent or protect the worker from the falling risk. [This opens up another avenue of questions and answers!]
- *If yes, how did you assess the likelihood of occurrence and severity of consequence?*
  - o Why did you assess the likelihood as 'Low'? [Witness must explain the logic or statistics behind the entry. Assessors usually get trapped here because many (if not most) simply follow previous years' entries, or use their personal or company's experience for the assessment. – They should use the industry's current statistics.]
  - o Why did you assess the severity as 'High'? [Witness must remember to include the controls that reduce severity but not those that do not reduce severity. For instance, a guard-rail reduces only the likelihood of falling to very low, but

does not affect the severity of injury once the person falls for one reason or another.]

- *In recommending controls for this fresh risk, did you follow the hierarchy of controls?* (Fig. 5.3.)

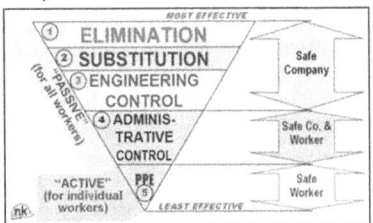

Fig. 5.3 Hierarchy of Controls

o Many sincerely believe that giving the worker Personal Protective Equipment (PPE) will safeguard him* from all harm.

[* Use of male pronoun will automatically cover the female counterpart, except where the context is gender-specific. – NK]

o But experiment and experience have shown the wisdom of following the hierarchy (order of decreasing effectiveness) to reduce injury.

o In the hierarchy, elimination of risk is considered first, followed by substitution with a less hazardous material, process, tool etc., engineering controls, and administrative controls, before providing PPE.

Obviously, a 'No', wrong or inadequate response to any of the above questions confirms a violation of legal norms and leaves the door open for penalties.

It therefore behoves a forensic engineer to be thorough in the Code of Practice for risk management [5.6], and related regulations. Author will present a few case studies to demonstrate the interaction between forensic engineering and risk management.

5.7 RESONANCE OF SUSPENSION BRIDGES

The Tacoma Narrows bridge in the State of Washington, USA, failed on 7 November 1940.

The Millennium Bridge in London had to be closed on opening

day, 10 June 2000.

5.7.1   Case Study 1 : Tacoma Narrows Bridge Collapse

In the case of collapse of Tacoma Narrows Bridge, called *"Galloping Gertie"* because of the way it failed, [5.7] the reason for the failure was a design flaw due to lack of knowledge on response of structures to aero-elastic flutter and vortex shedding, basically torsional oscillation, (Fig. 5.4).

Fig. 5.4 Tacoma Narrows Bridge Collapse

Although it was well known that soldiers marching over a flexible bridge caused high amplitude vertical oscillations due to resonance, the phenomenon of vortex shedding was not known at the time.

At that time also, the art and science of risk assessment as a strategy for elimination or mitigation of consequences of accidents was not sufficiently developed to have been of any use in anticipating or forestalling this disaster.

5.7.2   Case Study 2: Millennium Bridge Sway Problem

But the Millennium bridge [5.8], called the *"Wobbly Bridge"* because of the sideways pendulum-like motion, was designed and built by world-renowned designers and constructors (Fig. 5.5).

Fig. 5.5 Millennium Bridge

Yet, on opening day, when the people walked on to the just-opened bridge, the oscillations escalated to literally dizzying proportions to the extent it had to be closed. Subsequent analysis showed that the cause was 'positive feedback' phenomenon, known also as 'synchronous lateral excitation'. It took nearly two years and 5 million pounds to fix the problem.

But this problem was not unknown! While vertical resonance was known more than a century earlier, horizontal resonance had also been documented from the early 1970s for bridges with lateral frequency modes of less than 1.3 Hz, and sufficiently low mass.

A careful risk assessment would have identified the risk of such lateral motion on pedestrian bridges and appropriate pro-active controls could have been implemented to avert the subsequent embarrassment and expense.

5.8 NASA SHUTTLE DISASTERS

Can a world-famous best-endowed government agency be involved in two catastrophes of similar kind? Yes, the National Aeronautics and Space Administration of USA (NASA) had two similar problems in two of their space shuttles, one named *Challenger* [5.9] on 28 January 1986, and the other named *Columbia* [5.10] almost exactly 17 years later, on 1 February 2003, in each of which all seven of the crew died.

The circumstances under which the two shuttles were destroyed were quite different, as shown in Table 5.1, but there were certain common deficiencies in risk management. Ethical issues involved in these cases have been discussed by author elsewhere [5.11.]

Table 5.1 Comparison of Challenger and Columbia Disasters

Item	(a) Challenger	(b) Columbia
Date of disaster	28/1/1986	1/2/2003
No. of crew members dead	7	7
Main cause of failure	Freezing of 'O' rings	Impact of foam piece on wing
Time of failure	73 seconds after launch	During re-entry to earth
Role of Engineers	Warned about consequences of cold weather launch	Had warned about foam tile breaks; sought more information for rescue
Role of Management	Over-rode engineers recommendation and launched in cold weather	Ignored engineers' warnings and denied information sought

5.8.1   Case Study 3: Challenger Disaster

Data on failure probabilities at various low temperatures for the freezing of the 'O' ring sealing the connection for burning gases was available, as shown in Fig. 5.6, all predicting high probabilities of failure at temperatures below about 65°F.

Engineers had done a certain kind of risk assessment and concluded that with the overnight temperature of 18°F (–8°C) and even at the launch morning temperature of 28°F (–2.2°C) it was too risky to launch.

But the management out-talked and out-manoeuvred the engineers to the extent that key personnel changed their mind and approved the launch, with the disastrous results that followed.

Roger Boisjoly, an engineer who vehemently opposed the

launch in the face of heavy criticism was given an award for his 'whistle-blowing' on the violation of safety norms.

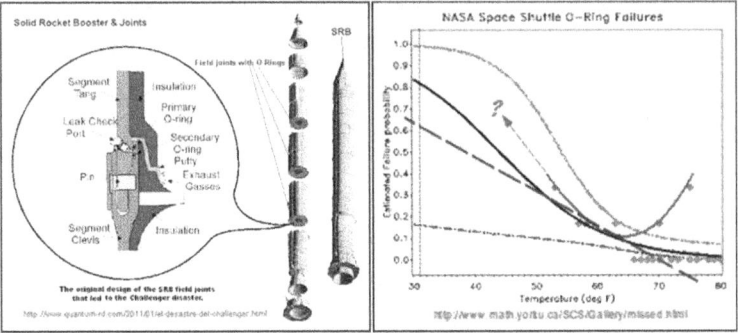

Fig. 5.6 Left: The 'O' ring that froze; Right: Various curve fits to failure

This is one glaring example where even with risk assessment pointing to high risk, managers could take action that might lead to disaster.

5.8.2   Case Study 4: Columbia Disaster

Columbia's history was even more colourful. During the launch on 16 January 2001, a piece of foam broke off the space shuttle's external tank and struck the left wing, (Fig. 5.7).

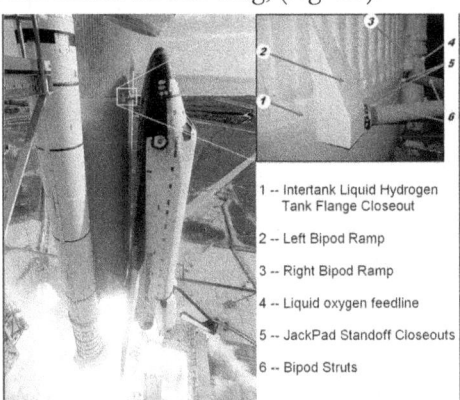

Fig. 5.7 Broken Foam Location

A few previous shuttle launches had seen minor damage from foam shedding with no significant consequences, and so nobody thought any more about it.

But when *Columbia* re-entered the earth's atmosphere on 1 February after completing its 16-day mission, the hole caused by the foam hit (travelling at Mach 2.46, i.e. 1,870 miles per hour or 840 meters per second) allowed hot atmospheric gases to penetrate and destroy the internal wing structure, which caused the spacecraft to become unstable and slowly break apart, burning itself and the crew on re-entry.

Here again, engineers were on the right track, wanting higher resolution images of the wing to assess the extent of the damage and work out some rectification control.

But the managers refused, influenced by the apparent futility of any such information.

As with the O-ring erosion problems of *Challenger*, NASA management became accustomed to the foam shedding phenomena in *Columbia* when no serious consequences resulted from these earlier episodes. Earlier warnings by engineers had also been ignored.

Apparently, the lessons from *Challenger* had not been learnt well enough, or the lessons learnt had been forgotten over the preceding 17 years. Risk assessment had pointed in the right directions, but the likelihood of disaster had been downgraded.

5.9 DESIGN CHANGES

Changes to design to enable fabrication or construction to take place are often carried out by the fabricators or constructors without the knowledge of the designers. Such changes can cause catastrophic repercussions, as will be illustrated by two classical cases, which the author had also presented at an earlier conference in a different context [5.12.]

5.9.1   Case Study 5 : Hartford Civic Center Collapse

Built in 1975 in Hartford, Connecticut, USA, the roof of the Hartford Civil Center, known as 'XL Center', collapsed early morning on 17 January 1978 due to a heavy snowstorm [5.13].

Only six hours earlier, 5000 had attended a basketball game. Luckily as the stadium was empty at the time of collapse, no one was injured.

The roof was 300 ft. by 360 ft.(90 m by 110 m), supported by a three-dimensional steel truss system assembled from pods 30 ft. by 30 ft. (Fig. 5.8).

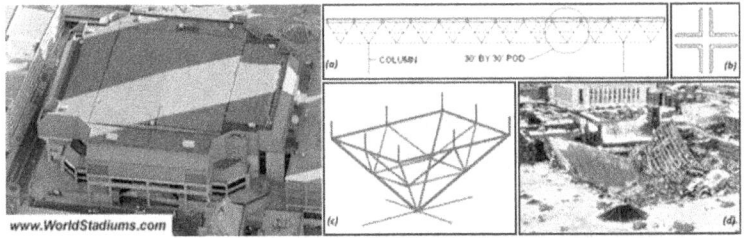

Fig.5.8 Left: XL Center; Right: (a) 3D Truss Section, (b) Member Cross-section, (c) Typical Pod, (d) Collapsed view

There were many reasons for the collapse, main ones among which were the following:

(1) Long compressions members were not braced in the middle as should have been.

(2) The computer analysis, being the first 3D truss analysis, was flawed.

(3) The contractor moved the connections a few centimeters (marked 'S' in Fig. 5.9) below the designed level for convenience of fabrication, resulting in reductions in capacity by as much as 90%.

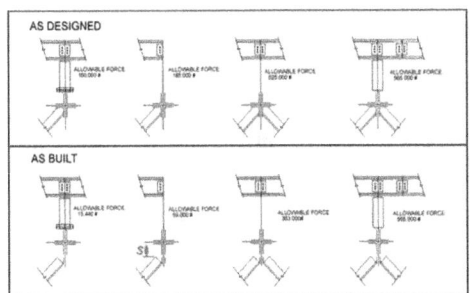

Fig. 5.9 Hartford Civic Center Roof Connection

(4) Warnings of excessive deflections from the site personnel during the erection process went unheeded by the designers.

The sad fact about the whole mess was that nobody bothered to assess the consequence of the changes made to the design and the possible reasons for the excessive deflections.

In other words, there was a complete lack of risk assessment and lack of risk control!

5.9.2   Case Study 6 : Hyatt Regency Walkway Collapse

Built in 1978, the second and fourth floor walkways of the Hyatt Regency Hotel in Kansas City, Missouri, USA failed catastrophically on the evening of 17 July 1981 [5.14], killing 114 and injuring 216 (Fig. 5.10).

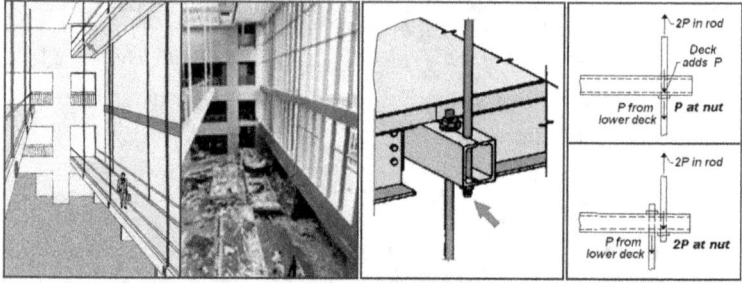

Fig. 5.10 Left pair: Hyatt Regency Walkway, before and after collapse; Middle: Nut at upper walkway; Right: Statics of single/double hangers

Although it had other shortcomings, the main reason for the collapse was the fact the fabricator changed the single rod design supporting both floors (as at top right in Fig. 10) to a double rod design each supporting one floor (as at bottom right in Fig. 10).

Apparently, approval was obtained for the change over a long-distance phone call.

The exact nature of the approval was under hot debate and court battle for a long time.

But what was not at all done was analysis of the actual effect of the change on the force distribution – in short, there was no risk assessment to check the result of the change. A static free-body analysis of the junction of the two rods at the third floor level would have shown that the nut there would be subjected to twice the load of the original design.

This situation addresses squarely the need for risk assessment and management in any task that requires re-design. Singapore has launched a thrust for integrating the safety in the design and construction phases, and in fact throughout the life cycle of the structure, by the newly mandated Design for Safety, to come into force from 1 August 2016.

5.10 CASE STUDY 7 : HURRICANE KATRINA LEVEE FAILURES

On 29 August 2005 Hurricane Katrina hit New Orleans, Louisiana, in Mississippi, USA, destroying more than 50 levees (i.e. embankments) and flood walls protecting the city and its suburbs (Fig. 5.11). [5.15.]

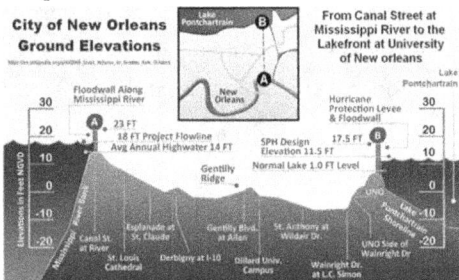

Fig. 5.11 New Orleans Ground Elevation

The levee and flood wall failures caused flooding in 80% of New Orleans and many townships in the vicinity. Tens of billions of gallons of water spilled into vast areas of New Orleans, flooding over 100,000 homes and businesses (Figure 5.12.)

More than 1200 people died in the hurricane and the subsequent floods, making it the deadliest USA hurricane since 1928. Total property damage was estimated at $108 billion (in 2005 USD).

Although the engineering failures were caused by a natural disaster beyond human control, forensic engineering highlighted many flaws in the emergency preparedness and crisis response system, as well as certain design and construction deficiencies in the levee system.

The main civil engineering deficiencies were:

Fig. 5.12 New Orleans, during Katrina & a decade later

- The project was incomplete even after 40 years of construction;
- Possibility of overtopping of levees and floodwalls;
- Designed pile depth below sea level was only 17 ft. (5.1 m), about half of what was required, due to a misinterpretation of pile test results in a cost-cutting exercise;
- Over-estimation of soil strength and hence under-design of levees leading to foundation failures;
- Inadequate pumping system and emergency power for it; and,
- Low factor of safety (1.3) in design, ignoring water gaps from older channels.

What is relevant to this paper is the inadequacy of the risk management system in the project. [5.16.] The risks of levee failure were never quantified before Hurricane Katrina.

Dam failure scenario was adopted for levee failure, which grossly underestimated the risk.

Likelihood of dam failure was set at once in 1 million years of operation, but likelihood of 1000 deaths from failure of hurricane protection system could be once in 40 years. Further the severity was also underestimated, as downstream evacuation after dam failure would have been relatively easy, while rescue from levee failure would be almost impossible.

As a result of this underestimation of risk, emergency preparedness was terribly flawed.

A risk assessment of areas that would be submerged would have indicated many vulnerabilities that were overlooked, such as escape routes, venues for emergency shelter, transportation facilities for evacuating people, etc.

The Superdome was the only place available to the displaced masses, and at one time 30,000 homeless people crowded in it without the basic amenities, for five days.

Even the National Guard headquarters were flooded. Figure 5.13 shows the hundreds of school buses which were expected to have been utilized for evacuation purposes, themselves stranded in the flood due to inadequate risk assessment and control.

Fig. 5.13 Rescue Buses Stranded in Flood

Risk management also failed in the communication of risks and controls to the main stakeholders who faced the risk, namely the public.

Due to lack of information and organization, there was complete chaos for a few days before the floods subsided, people were evacuated out of the Superdome, and some kind of order was restored.

It took five to ten years for the city to return to normalcy, with increased vibrancy.

5.11 PRO-ACTIVE FORENSIC CIVIL ENGINEERING

As has been proposed, deviating from the conventional definition of forensic engineering as confined to accident investigation, we may expand the term to cover the heuristic sense of engineering analysis carried out to eliminate accidents in proposed civil constructions and modifications, in such a way as to stand legal scrutiny.

Author would like to call this approach "Pro-active" forensic engineering.

Actually, such a definition would bring us squarely back to the sole aim of good design, namely to develop a procedure for developing a needed facility in such a way that it can serve with full functionality for a specified period of time without systemic or functional failure – in other words, satisfying the "Duty of Care".

Whether one is obligated to test a design under various failure scenarios may be a matter of debate, but as already mentioned, Singapore Government will require designers to check and 'design-out' potential risks in their designs from August 2016.

To address this and avoid accidents, risk assessment will become a pre-requisite.

The author would like to share two experiences involving his going beyond the call of duty:

(i) To explore what might happen if and when certain alterations were made to proposed procedures, and in that process some new risks were identified, leading to pro-actively preventing adverse consequences and failures from previously unidentified hazards; and,

(ii) To identify existing risks in a built structure and get them rectified.

5.11.1   Case Study 8 : Procedure to Install an Additional Sewer Line

Some time in the early 1970s while the author was in USA, his advice was sought in regard to the installation procedure for an additional steel sewer pipe as the existing brick-lined R.C. tunnel had reached its capacity.

Figure 5.14 illustrates the whole story, which author had also presented at an earlier conference in a different context [5.17.]

This was the era when the finite element method (FEM) of

structural analysis was getting into full swing, finding new applications every day. Author was among the first to use the method to analyse all kinds of artefacts from a filling in a tooth to a pre-tensioned steel bolted connection in his consultancy and research.

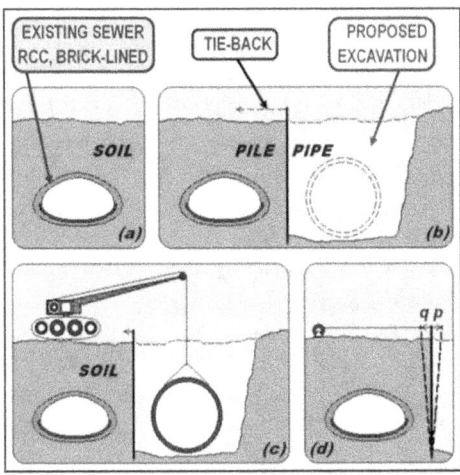

Fig. 5.14 Sewer Installation Risk Assessment

So he grabbed this opportunity to use FEM and in the next couple of days analysed a finite element model of the existing R.C. sewer, confirming that the extra load from the crane installing the new sewer in the excavation was safe, as long as the pipe-laying job was done slow and steady and steel plates were used to distribute the wheel loads.

As author had already developed the habit (or game?) of playing around with changing scenarios within the computer, he did a check on what would happen if the sheet pile tie-back slackened by 1/2 inch as at 'p' in Fig. 5.14(d), before the crane came on the soil bank.

There was no big problem with it and so he sent off his recommendation on Sunday evening that with due care they could install the new pipe as planned, on Monday morning.

During the night, he got the doubt what, if instead of slackening, an over-zealous operator tightened the tie-back 1/2 inch, as at 'q' in Fig. 5.14(d). He called his client and asked him not to start, and early next morning ran the computer analysis with displacement 'q' instead of 'p'.

He found that the crown of the old sewer would crack up under the lateral compression! Immediately he called back and advised the client not to install the new pipe from the old sewer (left) side, but somehow find a way to install it from the other (right) side of the excavation.

Author believes that it is this *"What if ...?"* approach to design, amounting to risk assessment and conforming to the new paradigm of 'Design for Safety', that would eliminate potential dangers in unusual structures and untested processes.

5.11.2 Case Study 9 : Risk Assessment of Handrail in Children's Park

This was one of the author's small contributions to safety in the township where he lived.

A few years ago, just having moved to his new address, he happened to notice that in the adjacent children's park, steps leading from the park down to the next level had inadequate edge protection. (Figure 5.15.)

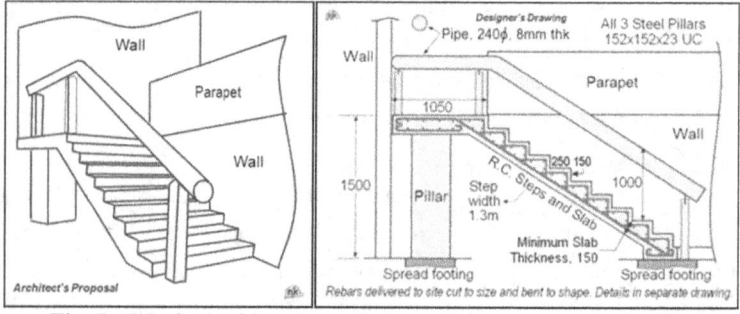

Fig. 5.15 Left: Architect's rendering; Right: Designer's drawing

He was able to get the problem rectified through the authorities.

Author had presented this at an earlier conference [5.12] in a different context.

As he was teaching a course on risk management, he later turned the situation into a tutorial on risk assessment and management.

He presented the problem as an architect's proposal together with an engineer's drawing (Fig. 5.15) and carried out a risk assessment on it, as in Table 5.2.

Table 5.2 Risk Assessment for Steps in a Children's Park

1. Sl. No.	2. Design Item	3. Hazards Identified	4. Risk – Consequences	5. Risk Assessment		
				Severity	*Likelihood*	*Risk*
1.	Handrail in use	Too large gap below handrail	Falling off the steps, injury	H	M	*H*
2.	Handrail in use	Top rail too big	No grip when needed, injury	H	M	*H*

Two hazards could immediately be identified during use of the steps, particularly by the children to whom the steps were intended, and by pregnant mothers and old grandparents accompanying them, as shown in Table 5.2.

For both the hazards, the risk, predictably, came up to 'High'. This meant that the task could not legally be carried out without reducing the high risk to at least 'Medium'.

Proposed control measures were a redesign with mid-rail and verticals, and reduction of hand-rail diameter to about 30-40 mm, as per the schematic shown in Fig. 5.16 and Table 5.3.

With these modifications, both risks would be reduced to medium.

Note that the severity remained unchanged at 'H', as the guard-rails reduce only the likelihood from 'Medium' to 'Low'. But the combination now reduced the risk to 'M'.

While this is a very simple, almost trivial, example, the impact of risk assessment on accident prevention is fully brought out through it.

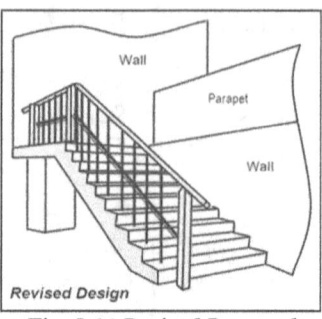

Fig. 5.16 Revised Proposal

Table 5.3 Recommended Redesign for Steps in Children's Park

1. Sl. No.	6. Can hazards be designed out?	7. Proposed Control Measures	8. Residual Risk			9. Further Review Required?	10. Staff & Timeframe
			Severity	Likelihood	Risk		
1.	Yes	Redesign with mid-rail and verticals	H	L	M	Yes, to confirm	AB 1 Month
2.	Yes	Reduce size of rails to about 30-40 mm	H	L`	M	Yes, to confirm	AB 1 Month

5.12 CASE STUDY 10 : FALL FROM HEIGHT INVESTIGATION

Author wishes to close with a recent accident investigation by authorities, findings and recommendations which might summarise the relationship between risk management and forensic engineering. In a recent seminar on fall from height accidents, it was reported that a worker tasked to install telecommunication equipment on a monopole (Fig. 5.17-a), fell off the ladder and died.

What follows is a review of the report by the enforcement and investigating authority. It was determined that the company had provided the requisite fall prevention device in the form of a 'rope-grab', a sliding handle which would allow the user to move along as long as the handle was kept pressed, but which would hold him up in case the user slips and loses his grip on the handle.

Fig. 5.17 Investigation of Fall from Height involving Risk Assessment

In Fig. 5.17-b, the rope-grab is shown enclosed in an oval, 'T' is the turnbuckle to maintain tension in the rope, and 'C' is the metal loop to which the worker must attach his safety harness.

The main reason the worker fell and died was that he had not attached his safety harness to the rope grab. This showed a lack of detail in the SWP and/or insufficient site supervision to ensure that the worker was using his PPE correctly and at all times.

Beyond this, the investigators discovered a violation which would have been adverse to the user even if he had been properly attached to the rope-grab. Normally the 'D' ring to which the harness lanyard should be attached for fall arrest is at the back, between the shoulder blades, as in Fig. 5.17-c.

However in the subject case, the user had attached the lanyard around the straps in front as in Fig. 5.17-d, which was a violation of the regulation, because if he had been attached to the rope-grab during the fall, he might have been strangled by the straps around his chest choking him at the neck.

1. Official findings by the authorities were reported as follows:

    (1.1) Risk assessment was not adequate for the installation work; and

    (1.2) Fall arrest system was not appropriate

2. Lessons learnt were reported as follows:

    (2.1) Risk was foreseeable;

    (2.2) Risk assessment was not conducted based on site

conditions;

    (2.3) Established risk controls were not relevant.

3. Summary of recommendations was as follows:

    (3.1) Hazards are often obvious;

    (3.2) Risk controls are usually common knowledge and practicable;

    (3.3) Risk assessment must be conducted on the site to ensure risk controls are relevant and effective.

Here, risk assessment should have been used to identify potential and credible dangers in the job, and then appropriate risk controls should have been documented in the SWP. Otherwise, violations of these and collateral controls (such as direct continuous supervision) will become the arguments for the prosecuting attorney to prove culpability of the accused.

This kind of interaction between risk management and forensic engineering is becoming an essential factor in meeting the requirements of the law.

5.13 CONCLUSION

Whether a forensic engineer chooses to become an expert witness or not, he is better off knowing the details of legal requirements concerning the accident he is investigating, particularly in regard to workplace risk management.

Explanations and examples provided in this paper should have conclusively confirmed the close relationship between risk management and forensic engineering.

An organization or a country that does not emphasise risk management at the workplace will not be able to extract the maximum information from its accident investigations, and by extrapolation, will not be able to reduce its accidents.

In another paper [5.18] author has explained the various uses of the risk matrix for risk management, many of which also may apply to forensic engineering.

While many lessons may be learned separately from the twin

topics of risk management and forensic engineering, their combined potential for knowledge on accident prevention and mitigation is greater than the sum of the individual contributions.

Forensic engineering is absolutely essential for risk assessment and control. It is now being realized that good risk management can not only cut down accidents but also serve as an investigative tool in forensic engineering.

A recent article in the *Forensic Engineering Newsletter* [5.19] states:

"Professional understanding among risk managers and forensic engineers has practical benefits. The two fields, ultimately, have similar professional focuses, function, purpose and work products. Recognizing this fact can only serve to help both professions."

The author can add that in his own experience, he has found risk management an indispensible ingredient for forensic civil engineering, whether to analyse an accident or to prevent an accident, particularly where legal implications may govern the outcomes.

5.14 REFERENCES

5.1.   _____, *On-line Etymology Dictionary*. Retrieved on 21 Nov. 2015 from:
http://www.etymonline.com/index.php?term=forensic

5.2.   Specter, M.M., 1987, National Academy of Forensic Engineers, *Journal of Performance of Constructed Facilities*, ASCE, 1, No. 3.

5.3.   Krishnamurthy, N., *Introduction to Risk Management*, (Self-Published), Singapore, May 2007, 88p, ISBN: 978-981-05-7924-1.

5.4.   _____, *Workplace Safety and Health Act*, Govt. of Singapore. Retrieved on 21 Nov. 2015 from:
http://statutes.agc.gov.sg/aol/search/display/view.w3p;que ry=DocId%3Aa7b4b808-d195-44ec-aa3d-dd5b1fa938f3%20Depth%3A0%20Status%3Ainforce;rec=0; whole=yes

5.5.   _____, *Risk Management Regulations*, Govt. of Singapore. Retrieved on 21 Nov. 2015 from: http://statutes.agc.gov.sg/aol/search/display/view.w3p;ord erBy=numUp;page=0;query=DocId%3Acd9437b7-419b-40de-99a3-09f2e7b8c90a%20Depth%3A0%20Status%3Ainforce;rec=0; whole=yes

5.6.   _____, *Code of Practice for Risk Management*, Workplace Safety and Health Council and Ministry of Manpower, Singapore, 2011 (2nd Revision 2015). Retrieved on 21 Nov. 2015 from: https://wshc.sg/files/wshc/upload/cms/file/CodeOfPracti ce_RiskManagement_SecondRevision.pdf

5.7.   _____, *Tacoma Narrows Bridge (1940)*. Retrieved on 24 Nov. 2015 from: https://en.wikipedia.org/wiki/Tacoma_Narrows_Bridge_(1 940)

5.8.   _____, *Millennium Bridge – London*. Retrieved on 24 Nov. 2015 from: https://en.wikipedia.org/wiki/Millennium_Bridge,_London

5.9.   _____, *Space Shuttle Challenger Disaster*. Retrieved on 24 Nov. 2015 from: https://en.wikipedia.org/wiki/Space_Shuttle_Challenger_di saster

5.10.   _____, *Space Shuttle Columbia Disaster*. Retrieved on 24 Nov. 2015 from: https://en.wikipedia.org/wiki/Space_Shuttle_Columbia_dis aster

5.11.   Krishnamurthy, N., "Forensic Engineering and Professional Ethics", *Proceedings of the Second International Conference and Exhibition on Forensic Civil Engineering*, Nagpur, India, 21-23 January 2016.

5.12.   Krishnamurthy, N., "Investigative Methods in Forensic Civil Engineering", *Proceedings of the (First) International Conference and Exhibition on Forensic Civil Engineering*, Bangalore, India, 23-24 August 2013.

5.13.   _____, *Building Collapse Cases/Hartford Civic Center*, Wiki. Retrieved on 24 Nov. 2015 from: http://matdl.org/failurecases/Building_Collapse_Cases/Har

tford_Civic_Center.html

5.14. _____, *Hyatt Regency Walkway Collapse*, Wikipedia. Retrieved on 24 Nov. 2015 from: https://en.wikipedia.org/wiki/Hyatt_Regency_walkway_coll apse

5.15. _____, *Hurricane Katrina*, Wikipedia. Retrieved on 1 Dec. 2015 from: https://en.wikipedia.org/wiki/Hurricane_Katrina

5.16. Delatte Jr., N.J, (Ed.), "New Orleans Hurricane Katrina Levee Failures", *Beyond Failure – Forensic Case Studies for Civil Engineers*, ASCE Press, 2009, pp. 287-299.

5.17. Krishnamurthy, N., "Use of Computers in Forensic Engineering", *Proceedings of the (First) International Conference and Exhibition on Forensic Civil Engineering*, Bangalore, India, 23-24 August 2013.

5.18. Krishnamurthy, N., "Construction productivity and risk management in Singapore", published in *'The Singapore Engineer' - The Magazine of the Institution of Engineers, Singapore*, February 2015, p. 22-29.

5.19. _____, "Risk Management from the Forensic Engineering Perspective – Part 2 of 2", *Forensic Engineering Newsletter*, Vol.2, Issue III, June 2007, Plick and Associates, Forensic Engineers, USA.

---

† This is a reprint of the author's paper presented at the Second International Forensic Civil Engineering Conference and Exhibition held at Nagpur, India, on 21-23 Jan. 2016, by the Association of Consulting Civil Engineers (India), reprinted with permission.

6

FORENSIC CIVIL ENGINEERING AND PROFESSIONAL ETHICS†

ABSTRACT

Professional ethics is considered by most engineers to be concerned more with philosophy of good conduct rather than with 'real' engineering. However, in many cases of civil (and other) engineering deficiencies, the root (or major contributory) cause is found to be failure in professional ethics in the design, construction/erection, or related administrative processes. In this paper, author attempts to bring together the two apparently disparate subjects into an integrated whole, using a number of case studies to illustrate his points, to demonstrates that by a proper overlay of the two areas, losses to society in general and the client in particular may be avoided, or at least their impact minimised.

6.1 INTRODUCTION

The subject of ethics is somewhat tough on engineers, because engineers are practical 'nuts and bolts', or (because we are talking about civil engineers) 'steel and concrete' folks who would rather be surveying in a desert or designing a skyscraper than wrack their heads about abstract and abstruse ideas of how ethically an engineer ought to behave.

But then, engineering is also the domain of noble achievements and sloppy jobs, and engineers should understand the difference between good and bad practice, and be able to draw a clear line between what is honourable and what is unprofessional.

Author, having taught Engineering Professionalism for years at the University level, understands what makes an ethical engineer, and with his immersion into forensic engineering, also realises the many ethical dilemmas that forensic engineers have to address, both at the level of tracking down unethical behaviour in their investigations, and at their own personal level, not to fall prey to temptation or expediency.

6.2 ENGINEERING PROFESSIONALISM

The classical theories of engineering ethics may be broadly divided into the following:

1. ***Virtue Ethics***
   - Propagated by Plato (c.428-c.348 B.C.) and his disciple Aristotle (384-322 B.C.)
2. ***Rights Ethics***
   - Propagated by John Locke, British Philosopher (1632-1704)
3. ***Universalist Ethics:***
   - Universalizability – Principle that an act is good if everyone should (may), in similar circumstances, do the same act without exception.
(3a) ***Duty Ethics:***
   - Respect for persons = Deontologism [*'Deon'*: Duty],

prescriptivism, propagated by German philosopher
Immanuel Kant (1724-1804)

(3b) Utilitarianism

- Teleologism [*Teleo*': End result], consequentialism,
  propagated by British Philosopher John Stuart Mill
  (1806-1873)
  (i) Cost-Benefit Analysis
  (ii) Act Utilitarian Approach
  (iii) Rule Utilitarian Approach

In the author's opinion and experience, virtue ethics has been
and will always be with us, and is more a benchmark to aim at rather
than a yardstick to measure by in practical terms. Virtues generally
do not impact forensic engineering, being more a moral imperative
than a legal constraint.

Duty Ethics and Rights Ethics are two sides of the same coin,
with the duty of A being a right of B and often vice-versa. They are
truly the stuff of forensic engineering, with courts always having to
decide whose duties were not carried out and whose rights were
violated. Most engineering Codes of Practice and standards
prescribe duties and rights of various stakeholders. Most Codes of
Ethics also rigorously insist on professional duties.

Utilitarianism or Consequentialism is the most commonly
applied theory in one or more of its variations in forensic
engineering. Forensic engineering is intimately concerned with
Utilitarianism in one form or another, with both sides of a case
invoking vigorous arguments to suit their view, and the judge having
to sort out which arguments of which side are more credible, and
more 'legal'.

(i) Cost-Benefit Ratio

is a measure of good over bad in a
material rather than in a philosophical sense, which is
applied in almost all modern endeavours with the over-
arching principle that any action is ethical only if the benefit
from it is more than its cost, both benefit and cost being
measured in dollars or other tangible form.

Trouble arises in two aspects, firstly in assessing the benefit from or cost of some consequence, and secondly in the choice of which stakeholders the benefit and cost should apply to. The frustrating part of this is firstly that often the larger public or environment is not included in the stakeholders, and secondly in many situations (including forensic engineering), even a life has to be – and is being – assessed in monetary terms.

(ii) ***Act Utilitarianism*** requires us to judge the ethical quality of an act on how much good it brought to how many people, and conversely on how little harm it brought to how few people.

It recognises the fact that in practice, in any act there will always be a few who will be harmed.

(iii) ***Rule Utilitarianism*** is the concept that human societies formulate rules of behaviour enforced by the authorities to maintain order and promote welfare among people. This is the easiest theory to apply, on the understanding that all members of a society must and will adhere to the rules made by their representatives or rulers, regardless of whether all the rules satisfy virtue or rights ethics.

Rules become the foundation of social duty ethics. A rule is called a 'Strong Rule' if it is applied without exception, and a 'Weak Rule' if exceptions are made depending on the merits of a case – too many of the latter reducing it to Act Utilitarianism.

6.3 LEGAL REQUIREMENTS OF PROFESSIONAL ETHICS

Most civilized countries have stringent requirements for the behavior of professionals in various fields, and engineering is no exception. Every branch of engineering has its own Code of Ethics.

Most commonly cited is the Code of Ethics by National Society of Professional Engineers (NSPE) [6.1]. With reference to forensic engineering, NSPE issued a position paper jointly with the National Academy of Forensic Engineers (NAFE) [6.2].

What is common to all codes and regulations governing forensic engineers are the following tenets ('Canons'), guided by the fact that the engineer's qualifications and testimony must stand up in a court of law, subject to the following considerations:

- Forensic engineering practitioners should limit their offering of services to the fields in which they have actual experience, or which may require only basic engineering knowledge.

- Forensic engineers should endeavour to provide objective, non-biased reporting and testimony, not slant them towards their client, and are obligated to report all findings, including those not favourable to their client.
  - o Expert witnesses do not win or lose a case; they only supply explanations and opinions regardless of which side of the argument pays them.

- Contingency fee compensation arrangements by Forensic Engineers are deemed to be unethical.

While violations of professional ethics in engineering practice often end up in the domain of professional societies and registration boards, leading to internal penalties and administrative strictures, when an accident happens, violations of professional ethics lead to court cases and stringent penalties, falling into the domain of forensic engineering.

Apart from the plaintiff and defendant being found to be unethical there is also the problem of the forensic engineer being unethical, which latter will be like a fox guarding the hen-house or like putting lunatics in charge of the asylum!

Author will present a number of case studies to illustrate determination of good and bad professional ethics through forensic civil engineering.

6.4 CASE STUDY 1: CITICORP CENTER

The Citicorp Center in New York (Fig. 6.1)was completed in 1977, with a very unusual design by LeMessurier (Inset in Fig. 6.1).

Fig. 6.1 Citicorp Center

Columns could not be placed at the four corners as in normal design (Fig. 6.2-a), because the owners of the St. Peter's Lutheran Church at one corner (marked with an oval in Fig. 6.2-a) refused to give it up for demolition.

Fig. 6.2 Citicorp Center (a) Conventional Design,
(b) LeMessurier Design, (c) As-Built

Citicorp made a deal to build them a new church in the same place if they would yield air rights above the church.

Based on this agreement, the corner columns were shifted to mid-side (Fig. 6.2-b), so that the building cantilevered 72 ft (22 m) over the church at one corner (Fig. 6.2-c).

In June 1978, a Princeton University student took up the Center for her thesis, and discovered that the building design was inadequate for a particular wind force [6.3]. By the time this question was communicated to LeMessurier, he had also found out that the design team, without his knowledge, had substituted the original welded connections with bolted connections, and further certain structural members which should have been treated as columns had been considered as truss members with more liberal design criterion.

On reviewing the design in July 1978, he found that the as-built design was under-designed by 40%, and subsequent wind-tunnel experiments confirmed this dangerous situation, which would have resulted in the toppling of the building under the anticipated wind gusts a few months hence.

Rather than stay silent (or commit suicide, which he contemplated briefly), LeMessurier not only presented the problem to Citicorp, but also offered them his solution of strengthening the bolted joints with welded gusset plates.

Had he failed to acknowledge the flaws in his design and/or rectify them promptly, the ensuing collapse would have cost thousands of lives, millions of dollars, and years of recovery.

LeMessurier was hailed as an 'Ethics Exemplar' – a role model for engineers faced with ethical dilemmas. He knew that disclosing the problem could lead to lawsuits and bankruptcy and possibly end his career, but he believed his selfish worries should take a back seat to his "social obligation."

His exhortation to engineering students in a MIT lecture was:

"If you've got a license from the State and a certification from the University first and now you're going to use the license to hold yourself out as a professional, you have a responsibility beyond yourself, if you see something that is a social risk… Good heavens, this thing would kill thousands! … You must do something. You must do something!" [6.3]

This was professional ethics at its best, and advice that every engineering student and every engineering practitioner would do well remember always and in all ways.

6.5 NASA SHUTTLE DISASTERS

The disasters that befell two of the space shuttles *Challenger* in

1986 and *Columbia* in 2003 launched by NASA, killing all seven crew of each have been described by the author in another paper [6.4]. Full information on them may be obtained from many sources [6.5, 6.6]. Table 6.1 from the author's paper is reproduced here for quick reference.

Table 6.1. Comparison of Challenger and Columbia Disasters

Item	(a) Challenger	(b) Columbia
Date of disaster	28/1/1986	1/2/2003
No. of crew members dead	7	7
Main cause of failure	Freezing of 'O' rings	Impact of foam piece on wing
Time of failure	73 seconds after launch	During re-entry to earth
Role of Engineers	Warned about consequences of cold weather launch	Had warned about foam tile breaks; sought more information for rescue
Role of Management	Over-rode engineers recommendation and launched in cold weather	Ignored engineers' warnings and denied information sought

Relevant to this paper is the fact that in both of them, professional ethics was involved, with management choosing to ignore the concerned engineers' recommendations and appeals.

6.5.1   Case Study 2 : Challenger Ethics

In the *Challenger* episode, engineers well knew in advance that launching the shuttle on such a cold morning would end in disaster, but the management – including some who also had engineering experience – chided them for over-reacting without adequate evidence, and lectured them to *"Take off your engineering hat and put on your management hat."* [6.7.]

That there had been temperature-related problems with the 'O' ring was known to all the personnel at NASA and its contractor Morton-Thiokol (Fig. 6.3).

The topic had been hotly debated many times, but it was the last-minute change from stopping the launch to recommending the launch by Morton-Thiokol under management pressure that allowed the launch to go ahead on the fateful morning.

The ethical issues involved were as follows:

Fig. 6.3 (Left) News Report, (Right) The 'O' Ring Problem

- One of the most common canons of professional Codes of ethics is that engineers shall *"hold paramount the safety, health, and welfare of the public in the performance of their professional duties."*
  o But the managers who over-ruled the engineers chose political expediency such as providing President Reagan a good punch-line for his impending State of the Union message, and ensuring continued funding for the space effort, over concerns for the lives of the crew in the face of warnings from experts even if not substantiated by hard scientific evidence.
- Their rash act also resulted in wastage of public funds and abuse of public trust.

Dr. Diane Vaughan termed this behavior as 'Normalisation of Deviance' [6.8], defining it as:

"The gradual process through which unacceptable practice or standards become acceptable. As the deviant behavior is repeated without catastrophic results, it becomes the social norm for the organization."

The one silver lining to the cloud was the courageous action of Roger Boisjoly, (Fig. 6.4) an engineer for Morton-Thiokol, the main contractor for the shuttle's rocket booster. Some six months before the launch he had written a memo insisting that behaviour of the 'O' rings in cold weather demanded extensive review before further launches were planned.

The night before the launch, he argued to stop the launch but was over-ruled.

Fig. 6.4 Boisjoly

When he testified before the Challenger Commission and filed unsuccessful lawsuits against Thiokol and NASA, he was ostracized by some of his colleagues to such an extent that he resigned.

For his honesty and integrity leading up to and directly following the shuttle disaster, Boisjoly was given the Award for Scientific Freedom and Responsibility by the American Association for the Advancement of Science in 1988. He became another 'Ethics Exemplar'.

6.5.2   Case Study 3 : Columbia Ethics

As in the *Challenger* case, foam break-up had been experienced by the *Columbia* shuttle in many of the earlier launches but without any serious adverse consequence (Fig. 6,5).

Hence, while engineers continued to worry and request reviews and further investigations, the launches went on, with the same behaviours as in the Challenger disaster namely 'Normalisation of Deviance', recurring.

Not only did NASA launch Columbia despite the reservations of the engineers, they also committed two ethical offences as follows:

(i) The engineers wanted high resolution images of the wing to examine and assess the damage to it, and devise whatever repair of the damage or rescue of the crew was possible. But their request was denied on the grounds that it would be a waste of time and resources which should be deployed elsewhere.

(ii) The management would not even inform the crew of the seriousness of the damage and its consequence of certain destruction of the shuttle and death of the crew on re-entry, on the grounds that as there was nothing that could be done to save the craft and the crew, it would be better for them not to know their impending death.

Fig. 6.5 (Left) News, (Right) Foam Hit

These two decisions again flew in the face of canons in most Codes of Ethics, which exhorted engineers to value human life and welfare above all else.

Management failed in its duty to do everything possible to ensure safety of the crew. It also denied the right of the condemned to learn details of their fate.

6.6 CASE STUDY 4: BOSTON TUNNEL CEILING COLLAPSE

On 10 July 2006 concrete ceiling panels (weighing tons) in *Big Dig'* tunnel in Boston, Massachusetts, USA, collapsed (Fig. 6.6) on the car of Mr. and Mrs. Del Valle (Fig. 6.7), on their way to celebrate their wedding anniversary, killing Mrs. Melina Del Valle instantly.

Fig. 6.6 Boston Tunnel Ceiling Collapse

Fig. 6.7 Mr. & Mrs. Del Valle

Major cause of the collapse was "epoxy creep" in the joint of the ceiling support screws to the roof.

Investigations revealed the following:

- Company which supplied the ceiling anchoring system did not emphasise that the epoxy was susceptible to creep and unsuitable for long-term load bearing (Fig. 6.8);

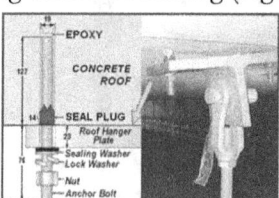

Fig. 6.8 Failed Anchor

- It did not identify anchor creep discovered in 1999; and
- The problem had been suspected earlier.

Ethical issues involved included the following [6.6,6.9]:

- A warning memo by Engineer Keaveney issued seven years earlier on the vulnerability of the epoxy glued joints was ignored.

- There was collusion between the main contractor and the State government in the matter of incentives. In particular the main contractor benefitted with an assured 7% profit on all cost overruns regardless of whose fault they were, as reward for the contractor hiding other excessive costs from the federal government.
- Corruption at high levels covered up the mistakes and diverted responsibility.

In addition to more than $500 million penalties paid by the main contractors to the state government, the Del Valle family received $21 million in compensation.

6.7 CASE STUDY 5: DENVER INTERNATIONAL AIRPORT

Planned in early 1990s, Denver International Airport (DIA) project faced many problems, (Fig. 6.9.)

Fig. 6.9 Denver International Airport

Although no actual accident occurred due to civil engineering misdemeanours, cracks appeared on the runway due to mistakes in the concrete, and the potential for major disaster existed, justifying the inclusion of this case study in this paper.

The most famous was the baggage handling system that turned out to be such a mess that it became a joke.

It took years to fix the various hardware and computer software problems.

In the civil engineering area, problems with runway and apron concrete made news in 1993.

Paving contractor Ball, Ball, & Brosmore ('3Bs') was sued for reducing the cement content of the concrete mix, saving money but weakening the concrete. [6.10.]

Charges on 3Bs included the following:

- Falsified inspection reports during construction.
- Falsified laboratory test reports, changes being justified later as 'engineering judgement'
- Inserted large (10 in.) clay balls inside the runway concrete to reduce cement and aggregate content.
- Altered actual material supply receipts to reflect the theoretically required quantities.
- 'Fool'ed the computer at batching plant to give correct measures by entering wrong data.
- Increased the cement content to meet specs during inspection, with prior notice of inspections from secret sources.
- Offered a concrete inspector a well-paid job to quit inspection.

3Bs settled claims of $300,000 for charges of falsification and unsound practice and $130,000 to two whistle-blowers.

The city of Denver chose not to sue the 3Bs or demand replacement of the shoddy work, as that would have caused more delays and problems.

Instead, they withheld $2.3 million for work that did not meet standards – a small fraction of the $138 million the city had paid the firm.

Critics were silenced with the assurance by 3Bs and some from the airport authorities that the lowered concrete quality would not appreciably affect service life.

But in 2006, two runways built by 3Bs were found to have deteriorated to an unacceptable danger level, and had to be replaced at a cost of over $38 million.

6.8 CASE STUDY 6 : TESTWELL LAB CONCRETE TESTING FRAUD

Testwell Laboratories in New York and its President Reddy Kancharla, were charged in October 2008 [6.11] with falsifying thousands of required concrete strength test results in connection with construction at Yankee Stadium and other major public works projects in New York.

Kancharla and three of his employees went on trial in December 2009, for concocting results of tests never performed and using computer projections instead of mixing and testing concrete in the lab. (Figure 6.10.)

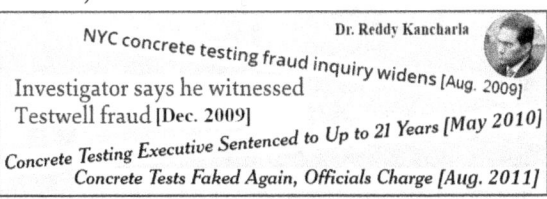

Fig. 6.10 News Headlines on Testwell Labs. fraud and CEO Dr.Kancharla

A former Testwell lab director testified that Kancharla trained him and other employees to falsify data. Workers were instructed to *"use a code word"* in e-mails about altering test results, court papers charge.

In his defence, Kancharla said that the mix designs were based on standard formulas and that clients knew what they were getting, implying that the test results did not influence the outcome.

Lawyer for one of his co-defendants claimed that as the practice was standard in the industry, his clients had not committed any crime.

After the trial Kancharla allegedly attempted twice to commit suicide, once by cutting his wrists and a second time by hanging himself, but failed both times.

In May 2010 Kancharla was sentenced to seven to 21 years behind bars and was ordered to pay $225,000 in reparations [6.11]. Testwell was ordered to pay $1.8 million in reparations.

Unfortunately, the story does not end there. American Standard

Testing and Consulting Laboratories, the company that the city selected as a replacement for Testwell Laboratories after the latter's indictment in 2010, allegedly had been doing the same kind of fraud on an even larger scale, and worse yet, they continued to do it with the freshly transferred projects also.

Not one of the 3000 test reports they submitted on the ongoing projects included any actual test results, but only fabricated values. The owner and five of his employees were charged with testing fraud on 4 Aug. 2011 [6.12],

Although again, no actual structural failure resulted from this fraud, scores of structures in which the concrete tests were involved were immediately inspected, cores taken, and their strengths checked.

In many cases, the actual strength was less than the spurious values cited by Testwell, but it was determined that there would be no immediate threat to structural safety, and only the expected life span may be reduced by about 50%.

Top officials of the company also manipulated government programs to obtain jobs for which they were otherwise ineligible.

lan Fortich, owner of American Standard Testing and Consulting Laboratories was sentenced on 14 Dec. 2012, for one to three years prison after pleading guilty to enterprise corruption [6.13].

The other five officials with the company which had defrauded the government of millions in fees with faked results and inspections regarding the strength and quality of construction concrete over the course of ten years, were to serve combinations of probation and community service for their supporting roles in Fortich's corner-cutting.

6.9 CASE STUDY 7 : HYATT REGENCY WALKWAY COLLAPSE

The catastrophic failure of the second and fourth floor walkways ('skywalks') of the Hyatt Regency Hotel in Kansas City, Missouri,

USA, on the evening of 17 July 1981, killing 114 and injuring 216 more have been described in another paper by the same author [6.4].

The change of design from a single rod supporting two walkways to two separate rods each supporting one walkway effectively doubled the load on the middle nut and led to failure. An overview of the case is provided in Fig. 6.11. Principal engineers were Gillum Associates.

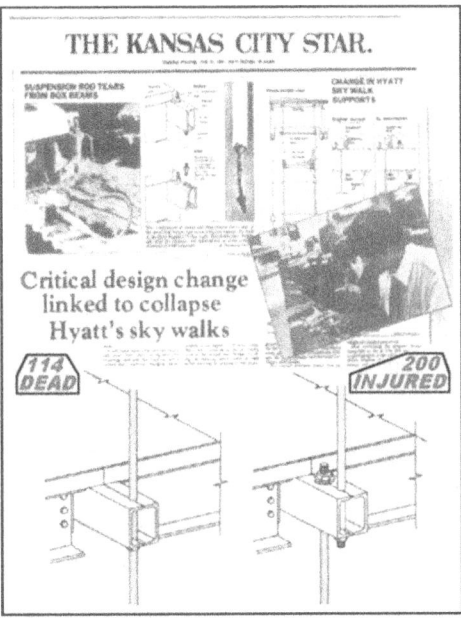

Fig. 6.11 Hyatt Regency Disaster

That case became unique in the history of structural failures because it raised questions on the validity of the Code of Ethics prevalent at the time [6.14].

The ASCE Code of Ethics which had been in force from 1914 had continued with minor changes until 1971 when an accumulating series of scandals, criticisms, and legal changes motivated ASCE to review their Code of Ethics.

The Task Committee identified the deficiencies in the existing Code as follows:

- It does not express concern for the public.
- It deals heavily with the 'business' aspects of engineering.

- It takes the negative approach by stating what you should not do rather than the positive approach of what you should do.

The revised Code, adopted in 1976, was based on the following three principles, hierarchically in the order listed:

1. Public welfare should be the primary concern of engineers.
2. The engineer has a duty to his clients, employer, or employees.
3. The engineer must be loyal to his profession.

The earlier Code assumed that an engineer was ethical until he was found to be otherwise. The revised Code declared that an engineer would be unethical unless he fulfilled certain fundamental obligations.

The entire philosophy of the Code was reflected as the Fundamental Canon Number One:

"Engineers shall hold paramount, the safety, health, and welfare of the public in the performance of their professional duties."

This became the hallmark of almost all the Codes of Ethics adopted in the civilized world from that time.

Subsequent modifications have reflected evolving changes in professional and social attitudes. For instance, a 2006 revision of Canon 1 reads thus:

"Engineers shall hold paramount the safety, health and welfare of the public and shall strive to comply with the principles of sustainable development in the performance of their professional duties."

However, the new Code was not without its opponents.

A strong objection was that putting public safety first would compromise other interests, and make the engineer liable for unforeseeable environmental and social impacts of the engineer's work.

The Hyatt Regency disaster became a test case for the new Code.

If the Gillum Associates' engineers Gillum and Duncan had merely overlooked a flaw and the accident had happened, it could have been written off as a costly but innocent mistake.

But in the Hyatt case, the faulty connection had been discussed no less than six times, before the accident. This outcome was pointed out as due to deficient risk management, as discussed in the other paper by the author [6.4].

Whenever asked, Duncan had answered that the revision was acceptable, and when confronted with the failure, he responded that the connection had not been designed by their company anyway.

Gillum had placed his seal on the revised connection (without checking its validity) and had also declared that "the entire atrium design" had been checked after an earlier accident, while actually the skywalks had not been checked at all.

The P.E. Licensing Board, after initial hesitation, broke tradition and filed a negligence suit against Gillum and Duncan. The two engineers claimed that because of the unfeasibility of any engineer being able to check all the details of any structure, Professional Engineers:

"routinely sealed and signed plans they had not personally checked" – as seems to be the practice in some situations all over the world even today!

They argued that they were merely members of the construction team and not 'design leaders' and hence could not be held individually responsible for the failure.

The judge pointed out that design engineers had a higher responsibility than fabricators or labourers, and had to accept responsibility for the safety of the user public in their structures. He found both engineers guilty of gross negligence and misconduct. Their licenses were revoked.

ASCE, in this first test of their ethics Code, chose to take a softer approach than expelling ASCE member Gillum from the Society.

They absolved him of the charges of negligence and unprofessional conduct, and simply suspended him for three years – but he resigned anyway.

Still facing engineers who wish to stop with disciplining but not punishing their colleagues who make deadly mistakes is the dilemma of how to deal with the public who suffer deaths of innocents and grave property damage from these mistakes.

At the legal level, the dilemma is resolved automatically by the courts deciding the culpability issues based on charges of injuries and damage brought by the aggrieved parties and assigning responsibility to all stakeholders responsible for:

(a) Creating the hazard and risk in the first place;

(b) Failing to develop and implement controls for the various risks; and,

(c) Failing to ensure that the person who faces the risk adopts the controls.

6.10 CASE STUDY 8 : VICE-PRESIDENT AGNEW'S BRIBERY SCANDAL

Spiro Theodore Agnew (1918 - 1996) was the 39th Vice President of the United States (1969–1973), serving under President Richard Nixon (Fig. 6.12).

Fig. 6.12 Agnew and Nixon

Son of a Greek immigrant, he rose from very humble circumstances through various elected offices to the Vice-Presidency.

In 1973, Agnew was investigated by the United States Attorney for the District of Maryland on charges of extortion, tax fraud, bribery, and conspiracy [6.15]. He was charged with having accepted bribes totalling more than $100,000 while holding office as Baltimore County Executive, Governor of Maryland, and Vice President.

The Baltimore Sun reported [6.16]:

"Agnew's criminality was straightforward: I'll see that you get a lucrative engineering contract; you give me 5 percent in cash, and we'll

both be happy."

In addition to straight pay-offs, various schemes were hatched to conceal money still being paid after Agnew left the State House, including as "legal fees" to be paid after Agnew left office or as "loans."

With two other highly placed engineers, Agnew split large bribes three ways with him taking 50 percent, and the other two getting 25 percent each.

Engineer Green who while Agnew was Maryland State Governor got a lot of State work, paid him more than $20,000 a year, and continued to pay somewhat less even after Agnew became Vice-President, until the latter's indictment.

It was only when engineer Matz was caught in a bribery case (not involving Agnew directly) that he revealed his payments to Agnew. Then, the whole sordid mess spilled out, and Agnew was formally charged in 1973.

But Agnew made a "plea bargain", admitting to a tax evasion charge and resigned, thus escaping being convicted for bribery.

He resigned on 10 Oct. 1973, paying a fine of $10,000 toward tax evasion, and being slapped a three-year probation. He also "reimbursed" the State $268,000, the bribes he was estimated to have collected during his tenure (Fig. 6.13).

Even then, he wouldn't give up.

He tried to take a tax deduction for his reimbursement, but that was disallowed!

Although in this case study there were no accidents, injuries, or fatalities, there were numerous ethical misdemeanours to be investigated, involving the loss of considerable funds to the government, and ultimately a plundering of public taxes and trust.

Nixon himself was to be impeached for his involvement in burglarizing Democratic party headquarters at Watergate in 1972. He resigned in 1974.

Fig. 6.13 Agnew's Resignation

6.11 ETHICS OF FORENSIC ENGINEERS

As in most other human activity involving power and money, there will be at least a few forensic engineers who may try to use unethical means to achieve their ends, in particular to slant their findings in favour of their clients or other benefactors.

This possibility is professionally and socially harmful because it will be like a policeman robbing a tourist. The forensic engineer has at his command a spectrum of skills and tools to investigate an accident and by the same token has also the capability to misuse the same skills and tools to mislead the opponent or even to take advantage of his own client.

Expert witnesses must be especially on their guard that they do not suppress vital information which may hurt their side or benefit the opposite side. They should also not agree to contingency fee arrangements whereby clients or attorneys would stipulate that they would be paid a more handsome fee if their clients won the case. Expert witnesses discharged from a case before its closure, must not switch sides.

However, there is a natural check and balance in this, because the same law that they try to circumvent will sooner or later catch up with them, and deal with them in ways harsher than it would a lay person in society – somewhat like a lying lawyer will be ostracized by peers and be disgraced in public losing livelihood and hard-won reputation.

6.12 CONCLUSION

Professional ethics governs the correct behaviour of engineers. Forensic engineering is a matter of investigating accidents, but in its expanded definition of legal implications, prevention of accidents by legal means via ethics is also a professional imperative.

To that extent, study of previous accidents with well-documented case-histories will bring home to engineering students and practitioners alike, the fact that fully ethical behaviour will not only make the profession much nobler, but also assure the public that they are getting the best, safest, and most economical structure, facility or service possible for their money and effort.

Engineers in general and civil engineers in particular, are the implementers and facilitators for any society in its aims to provide its citizens with all the structures, facilities and services that they need to implement and utilize all their resources and fulfil all their aspirations in the social, artistic, cultural, scientific, and technological domains.

Placing ourselves on this high pedestal, sheer pride and self-respect of being an engineer should keep the few who fall short from pulling the silent and ethical majority from tumbling down!

6.13 REFERENCES

6.1. _____, *Code of Ethics for Engineers*, National Society of Professional Engineers, USA, Revised July 2007.

6.2. _____, *NSPE Position Statement No. 1748--NSPE-NAFE Joint Position on Forensic Engineering.*

6.3. Vardaro, M.J., *LeMessurier Stands Tall – A Case Study in Professional Ethics*, AIA Trust, USA. Retrieved on 28 Nov. 2015, from:

http://www.theaiatrust.com/whitepapers/ethics/study.php

6.4. Krishnamurthy, N., Forensic Engineering and Risk Management, *Proceedings of the Second International Conference and Exhibition on Forensic Civil Engineering*, Nagpur, India, 21-23 January 2016.

6.5. _____, *Space Shuttle Challenger Disaster*, Wikipedia, Retrieved on 24 Nov. 2015 from:
https://en.wikipedia.org/wiki/Space_Shuttle_Challenger_dis aster

6.6. _____, *Space Shuttle Columbia Disaster*, Wikipedia, Retrieved on 24 Nov. 2015 from:
https://en.wikipedia.org/wiki/Space_Shuttle_Columbia_disa ster

6.7. _____, The Space Shuttle Challenger Disaster, *The Engineer*, 24 Oct. 2006. Retrieved on 28 Nov. 2006 from:
http://www.engineering.com/Library/ArticlesPage/tabid/85 /ArticleID/170/The-Space-Shuttle-Challenger-Disaster.aspx

6.8. Roe, Rob, "The Normalization of Deviance (If It Can Happen to NASA, It Can Happen to You)" , *On the Line – Public safety Risk Management*, 28 Jan. 2013. Retrieved on 28 Nov. 2015 from:
http://lmcontheline.blogspot.sg/2013/01/the-normalization-of-deviance-if-it-can.html

6.9. _____, *Professionalism/Boston's Big Dig Project*, Wiki Books. Retrieved on 28 Nov. 2015 from:
https://en.wikibooks.org/wiki/Professionalism/Boston%27s _Big_Dig_Project.

6.10. KIlzer, L., and A. Gathright, "Finding concrete problems at Denver International Airport", *The Denver (CO) Rocky Mountain News*, July 15, 2006. Retrieved on 24 Dec. 2015 from:
http://archives.californiaaviation.org/airport/msg37847.html

6.11. *Testwell Laboratories owner and 3 employees go to trial*. Retrieved on 30 Nov. 2015 from:
http://www.aggregateresearch.com/articles/18014/Testwell-Laboratories-owner-and-3-employees-go-to-trial.aspx

6.12. Rashbaum, W.K., "Concrete Tests Faked Again, Officials Charge", *The New York Times*, 4 Aug. 2011. Retrieved on 29 Nov. 2015 from:

http://www.nytimes.com/2011/08/05/nyregion/6-charged-with-falsifying-concrete-testing-results.html

6.13. Jacobs, Shayna, *Concrete Testing Boss Gets Prison Time for Faking Test Results.* Retrieved on 29 Nov. 2015 from: http://www.dnainfo.com/new-york/20100526/manhattan/concrete-testing-boss-gets-prison-time-for-faking-test-results

6.14. Pfatteicher, S.K.A., "Walkways: Tragedy and Transformation in Kansas City", *Proceedings of the Second Congress in Forensic Engineering,* (Ed. Rens, Rendon-Herrero and Bosela), ASCE, San Juan, Puerto Rico. 21-23 May, 2000, pp.47-56.

6.15. _____, *Spiro Agnew,* Wikipedia. Retrieved on 30 Nov. 2015 from: https://en.wikipedia.org/wiki/Spiro_Agnew

6.16. _____, "Agnew took hundreds of thousands of dollars in contractors' kickbacks Envelopes filled with cash given to him in his office detailed in 40 page report", *The Baltimore Sun,* 19 Sep. 1996. Retrieved on 30 Nov. 2015 from: http://articles.baltimoresun.com/1996-09-19/news/1996263146_1_agnew-white-envelopes-baltimore-county

---

† This is a reprint of the author's paper presented at the Second International Forensic Civil Engineering Conference and Exhibition held at Nagpur, India, on 21-23 Jan. 2016, by the Association of Consulting Civil Engineers (India), reprinted with permission.

ABOUT THE AUTHOR

Dr. N. Krishnamurthy, known as 'Prof Krishna' to his students and peers, has more than five and a half decades of teaching, research, and consultancy experience in structural engineering, computer applications, and workplace safety and risk management.

He has authored four books and co-authored another four, and published about hundred papers.

Relevant to this book of essays are that he has taught courses related to accident investigation and prevention in the National University of Singapore and the University of New South Wales (Australia) – Singapore campus, and short courses for the Institution of Engineers (Singapore). He has consulted for and carried out accident investigations for the Ministry of Manpower, Singapore, and for private parties in USA and Singapore.

More detailed information on his academic and professional background may be found at his website: www.profkrishna.com

———